Gary Hays

Well Born

For Mark,
Bonita,
Bobby,
Catherine,
Kirk,
and Danny

FORWARD

WELL BORN is in part a work of historical fiction. Hundreds of published materials on the eugenics movement were referenced over the course of writing this novel. The author's research included a thorough review of the personal papers of Harry H. Laughlin, an early leader in the eugenics movement in the United States. Those papers are maintained in the Pickler Memorial Library at Truman State University in Kirksville, MO.

As a work of fiction, **Well Born** includes incidents and dialogue, as well as characters that are the product of the author's imagination. Where real life characters appear, the situations, incidents and dialogue concerning those persons are entirely fictional and are not intended to depict actual events, or to change the entirely fictional nature of this work. In all other respects, any resemblance to actual persons, living or dead, or locales is entirely coincidental.

CONTENTS

PART I
NIGEL AND EUGENE

PROLOGUE

Nigel Wellbourne was a thinking man. In that, he was not like most men, men from whom he differed *in extremis*. Those wankers, those clods were men of less refined sensibilities, they who impetuously strove to fulfill desires of the carnal kind, they who too readily neglected reason, with nary a consideration for the scientific or mathematical. They were to him the rabble mobs, the crass and the disreputable, tenaciously clinging to staves and religiosity, bitterly holding back against the progressive arc of rational thought. Not our young Wellbourne. Nay, Wellbourne's *raison d'etre* was of a higher order. With a propensity to ponder the imponderable, particularly on the nature of Virtue, his thoughts, of necessity, occasionally succumbed to more immediate human concerns for gainful employment. Whilst achievement of Virtue was not what he imagined he might fully achieve in his own short life, and certainly not what he saw in the lives of the masses, it was a conception worthy of dedicated contemplation in and of itself, and thus worthy of his considerable time and devotion.

[Oh, begging your forgiveness; I've gotten a tad ahead of myself. Perhaps you may not trust in my observations about Wellbourne. Allow me to qualify my right to this narration by saying a bit about my familiarity with him. I suppose you might say I have known him forever. It could rightly be said that I knew him better than he knew himself. Indeed, we have spent sufficient time

3

together that I knew his innermost thoughts and the words he was to speak before they were ever formed on his tongue. It is true that I knew him even before he was knit in his mother's womb. Nigel Wellbourne hid no secrets from me. Now, where was I? Ah, yes … right.]

This man Wellbourne, even at a young age, brought a disciplined logic and mathematical rigor to his meditations. It was, of course, the manner in which he had matriculated, where at the University College of London he naturally read in math and philosophy, and studied the great thinkers of his time. Such learned men as Sir Francis Galton, Thomas Malthus, Charles Darwin, and Karl Pearson had an outsized influence on him at his most formative time. They were among the intellectual establishment, and were, if only from a distance, Wellbourne's guides and mentors to a newly acquired life of the mind. It had been their reasoned arguments on the present lot of man, and the means of improving the human condition, of which he had read repeatedly, that taught him to vigorously challenge the most tenaciously held of his once more conventional views — not to mention those of the *hoi polloi*. For these progressive men sought a higher order, even in man as a species.

Later, as a first-year seminarian at Wycliffe Hall, Wellbourne was to hear many celebrated authorities lecture on topics of great moral significance. Fortuitously, perhaps even prophetically, he was present for a talk by the esteemed though superannuated Right Bishop F. Barclay Nuttal. It was Nuttal's biennial lecture on the Pauline Epistles that had aroused great passion in Wellbourne.

[Permit me to interject. It is worthy of emphasis that Wellbourne, though outwardly reserved, was vulnerable to passion in the same way as Paul. When well directed, passion can be noble,

but it can also easily manifest in less than admirable pursuits, as our story will reveal.]

Nuttal's monologue, revered for its much-awaited terminus — if not its comprehensiveness — was crowned with a verbal flourish that had the audience gasping rapturously. As with the Apostle Paul, Nuttal entreated modern man to get his degenerate mind, even his whole being, out of the gutter.

[On this, I could not have agreed more.]

So it was that, through Nuttal's brilliant oratory, young Wellbourne's life course was set in motion after some initial delay. Throughout his life, Nigel was to keep an engraving of this most salient verse on his desk: "Finally, brothers and sisters, whatever is true, whatever is noble, whatever is right, whatever is pure, whatever is lovely, whatever is admirable--if anything is excellent or praiseworthy--think about such things." [Phillippians 4:8] So our man Wellbourne would think, and in due course venture where few men dared to go.

[Paul's was a sterling admonition but, one would have to agree, exceedingly hard to follow amidst the vicissitudes of life.]

(Vicissitudes, Padre? Throughout history, have you yourself not been at the nexus of countless vicissitudes? And while I'm at it, why must you insist on retelling this old story? Haven't you something more contemporary to share? Anyway, I would have thought you might have seized this occasion to introduce me.)

[Yes, of course — my apologies. Do please allow me to introduce my associate and interlocutor, whom I familiarly address as Consigliere — an appellation of course. He is my intercessor, if you will. We go way back. I expect he will insert, or should I say assert himself in this tale. He is known for nothing else if not his wise counsel — hence the cognomen.

As for a more up-to-date story, I will share one in due course. As you will see, they are in many ways the same story, and that I think may be largely the point, or at least an important point.]

It must emphatically be stated that Wellbourne was neither contemptuous of, nor inured to the plight of the masses. Were not all men, even such as himself, merely animals, a derivation, a variant fraught with all the attendant limitations of biological dead ends and dry branches in an evolutionary stream? Alas, it was his calling as a minister of God that permitted, no, required of him the need to balance compassion for the flock with this private striving for ascendency. Even as he was to minister in the most mundane matters of new life, ill-timed death, and ever-present discord — especially in matters of worship music and liturgy — our inveterate seeker secretly forged his own intellectual path. In the end, it was a balancing act that required more of Wellbourne's temperament than he was able to master. For in time, Wellbourne grew weary of the Anglican flock. It is true that his con-gregants' greatest concerns often centered on vigorously protecting or celebrating position, possessions, and privilege. But as Nigel saw it, the great unwashed too often suffered more from enfeeblement, insanity and moral turpitude. Toward the end of his religious sojourn, Nigel had grown restless, even contemptuous of the problems of the church. He thought they were much too trivial. He knew in his heart that something important must be done for the plight of the Malthusian masses, something beyond religious piety and soporific platitude.

[One more thought. If on occasion we seem insensitive, it is only that we have attempted to employ the vernacular and values of the times. Now, as to your comment on vicissitudes, does it follow that what gives rise to man's troubles is my mere passivity? Is not man's freedom commonly the source of his own troubles? In any

event, should I not judiciously wait for a propitious moment to intervene?]

(A point well-taken, and if I may add a word, the reader should be given to understand that the meaning of words, and in particular the emotions and values attached to them, may change over time in accord with shifting cultural norms. This should not, however, convey that some new form enlightenment has necessarily been attained. For in euphemism lies the wicked serpent of obfuscation. If considering words, and therefore ideas from a former time is impermissible, one cannot have a complete sense of the past, let alone chart a path out of the morass.)

[Thus begins a tale, which our point in sharing is, we hope, instructive, and even perhaps a little amusing. In the telling, we shall try mightily to avoid intrusion, but knowing ourselves as we do, and in accepting sometimes faltering memory borne of age, allow us to humbly beg indulgence if we cannot completely refrain from occasional commentary.]

CHAPTER I

WHEN WELLBOURNE COMPLETED HIS ORDINATION, he was posted to a small church in Letchworth, a village noted primarily for being the site of the country's first roundabout, the latter being located at the juncture of the only two roads through town. Letchworth was a place not unlike those typically assigned to new and somewhat less-than-promising ministers of God. It was in most respects quintessential England, at least of the early post-Victorian era. Its morés and customs, like vexatious habits, were unchanged for more than two hundred years.

It was good that he had not as yet found a mate. In fact, given his seeming outward lack of vitality and an overly abundant thought life, Wellbourne would not likely ever find a mate, despite his tall, lean physique, and prepossessing good looks. He appeared to most of his fellow men an odd duck, in spite of his pleasing features. Stiff, reserved, and bumbling in the social niceties, Nigel was not then a man who met people easily, particularly young women. Under no circumstances did he manage anything beyond perfunctory social relationships, but he concluded that no more was expected of him as an Anglican minister. Still, he was capable of empathy, even kindness, and in a subdued way manifested the full range of human emotions. And to the carnal desires of which we have earlier spoken, he was no complete stranger. It was just that, in the way of a mathematician, he thought it best to coldly and precisely hold these in check according to some self-imposed, mathematical order of operations.

Once Wellbourne had become sufficiently established in the church, he began to venture forth throughout Letchworth and its surrounding area. For those sojourns, he relied on a rickety old bone-shaker, which had been lent by an elderly parishioner. On one such venture to the local market, he chanced upon a comely maiden of approximately twenty years. They both had been waiting patiently in line for quince, a circumstance to which only Wellbourne might attach significance. Observing her struggle to make a selection with fully laden arms, Nigel dared to intercede with an offer to hold her basket. It was for him a gesture both risky and profoundly out of character. She smiled pleasantly, respectfully declined in what seemed to Wellbourne a most coquettish way, and they parted company with no further interaction. Doubtless he was at that instant smitten.

The following week, they happened upon each other again, and exchanged shy sideways glances. Over the ensuing weeks, they reached the point of amiable fraternity, and could at last acknowledge each other with verbal greetings. These encounters went on for months before their market meetings reached what for them was a sufficiently intimate level that they dared to move beyond a mere greeting. It was Gemma who made the first move, when she observed that she had seen Curate Wellbourne many times at the local church. Taken aback by her boldness, Wellbourne was so encouraged that, on recovering, he remarked, "Oh."

Some weeks later, Nigel summoned his courage once again, and asked her to join him on a bench in the town square. About the plan for this invitation, Wellbourne had ruminated during many sleepless nights. There, in the park in the center of town, a town that was in the center of the garden that was England, which many erroneously considered to be the center of the universe, they spoke to each other daringly, not just of fruit and weather, but of themselves. It was then, in a gesture of sublime magnanimity, she offered him a taste of her fruit —fully ripened no less — and they partook of it.

In time, Wellbourne learned that Gemma was an attendant to the Lady Burgess, a local widow of some considerable means. Gemma lived alone on the third floor of the lady's manor house. Her two small rooms were accessible via a back stairway off the kitchen. Though Gemma had inherited a small sum at her father's passing, with no remaining family she of necessity relied on the kindness of Lady Burgess for her sustenance. In exchange, she performed light house-keeping duties. It was in the performance of these duties that Gemma had occasion to encounter Mr. Wellbourne during weekly marketing forays. As might be expected, Gemma's exchanges with Wellbourne led to sharing their respective hopes and dreams, which in turn led to such intimacy that they eventually could sit side by side, and even hold hands. It was as yet no scandal that their innocent courtship, though unsupervised, in time gave rise to emotions that neither had thereto-fore suffered — hers to a noticeable and persistent pulsing at the temples, and his to a not unpleasant quickening in the loins.

(*Ought I to have intervened at this point in their courtship?*)

[**I think not, lest we might have obstructed revelation of true character. Be still now, and let me tell the story**].

One must confess that some of their encounters shaded the line of accepted church standards, it being vastly more common for new ministers of the church to have been wed by the time of their first assignment. Notwithstanding Gemma's and Nigel's attempts to keep their relationship a closely held secret, their budding relationship was a frequent topic of over-the-stile discussion. Indeed, Curate Wellbourne was, to local wenches and maidens alike, an eligible bachelor, whilst a threat to the layabouts and louts of Letchworth. In no small measure, his attractiveness to these lusty lasses rested more on his natural good looks than his perspicacity, for they were better equipped to discern the former than the latter. To the congregants, his expected betrothal made him a trustworthy and less threatening presence.

As their love grew, they were at last unable to restrain a passion in which lascivious desire did indeed breech the bounds of saintly righteousness, if not clerical standards of behavior. As with exponential growth, familiarity inevitably crept into love, and love, passion, only to burst forth geometrically into lust. And lust was to be a determined master at that time. Thus it was that they conspired to see their mutual attraction reach its fullest flower. It must be said that in later reflecting on their, their … what he referred to as a tête-à-tête, Wellbourne determined that he had been the passive one in the matter.

[While I know he considered Gemma to be the instigator, in both cases it had begun with the same thought.]

(*You have always cautioned about misguided thinking. A confession is warranted here, don't you think? It should have fallen to us, certainly to me, to redirect their thoughts.*)

The Lady Burgess had been planning a fortnight visit to a cousin on the Continent. Save for her customary housekeeping duties, Gemma would be free during that time. Thus, over a two-week period, Nigel would finally discover what it was that produced the evolutionary stream of which they both were now most decidedly a part. So, carry on they did. It should have therefore come as no surprise to them that, just as correlation is sometimes connected with causation, conjugation is not uncommonly linked with conception. Stunned and immobilized by fear, they endeavored to keep their secret for as long as possible. Thinking man that he was, our young Nigel, having attained twenty-seven years, had neither a plan A nor a plan B. And like other men of God, he too often called on Providence to work out the details.

[It never ceases to amaze me that, though the occasions for supplication are legion, men insist on self-reliance until things have gone all to pot. To paraphrase a scene from one of my favorite movies, "He never wanted my friendship. He was afraid to be in my debt. But now he comes to me and says, Godfather, give me justice. But he doesn't ask me with respect. He doesn't offer friendship. He doesn't even think to call me Godfather. What have I ever done for him to treat me so disrespectfully?"]

(Oh, please! Are you not able, for even just one moment, to desist from quoting some movie or other?)

It was obvious why Nigel had been so easily attracted to Gemma. Like Wellbourne, she was possessed of uncommon natural beauty, a trim athletic figure, and a countenance that invariably drew people in. A radiant smile with exceptionally straight white teeth, and sparkling blue eyes, made her one of the most eligible young women in Letchworth. Looks aside, unlike our Nigel, she had an engaging personality and could within moments acquire deep knowledge of those with whom she shared intercourse. Perhaps it was because she was capable of denying self, and assuming the condition of the other. Or perhaps it was a capacity to elicit more than a small amount of information through deft questioning, keen observation, and sympathetic listening. In any event, she could through native intelligence deduce much from those she encountered. In the beginning, it was this sagacity that enabled her to plumb the depths of her paramour's thoughts, and at last find something of which to take hold.

Fortunately for both Nigel and Gemma, Lady Burgess was of the most progressive sort. When she learned of Gemma's condition, she not only supported, but grew even closer to Gemma. Together they made plans for the baby, the Lady neither fully expecting nor particularly wanting that the father would be publicly identified.

Community intruded on her plans, however. Observing Nigel's and Gemma's poorly masked mutual affection, and subsequent changes in Gemma's physique, it became apparent to the Lady, and in time to the community that the young lass was indeed with child. A slight paunch, and unexpectedly excusing herself from Sunday services were two such indications. In time, tongues wagged — no, I dare say they lashed, and with animus. For it was not then expected that a minister of God shouldst impregnate a young woman to whom he was as yet un-betrothed.

It was inevitable that the church would take a stand. On most issues, the church had found it impertinent to assert itself, but given the hue and cry of parishioners, it was unavoidable.

(*Judge not, lest you be judged and all that.*)

[**Quite! Yet in this case ill-timed.**]

The church was neither accustomed to nor prepared for putting itself in a position of moral judgment, at least other than in the abstract. Pushed to act nonetheless, the vestry at once appointed a committee to ascertain the veracity of the charge. Delays ensued, by which time the assertions of the locals became undeniable. Forthwith, an *ad hoc* church probity committee was formed to confront Lady Burgess.

"Madam," a church elder said, "We beg your indulgence in what is a most delicate matter. It is our charge to inquire into the young lady with whom our curate is acquainted."

"You are asking about Gemma?" the Lady replied. "Pastor Wellbourne's lover? Well, aren't you the lot of nancies?" For a time, they were stunned into silence by her rude response. Looking down on them with a withering gaze, she asked, "What exactly is it you are wanting to know?"

"Emm, ah, this is most delicate," the eldest elder said, as his fellow committee members looked over his shoulder. "I should say, then, emmm, it's irregular, but a … er … emm, could it be that, uh, she is, uh …?"

"With child? Knapped?" the Lady replied, feigning shock. "Of course she is. You boys seem surprised. Just how in bloody Hell do you think it is that you all came to be?" Her manner was as usual direct, even daft. Annoyed at their timidity, the Lady pressed her lips tightly together and shook her head, stunned at the obtuseness arrayed before her. Hers was, to the committeemen, a stunningly rude expression in reply to what to them seemed a perfectly diplomatic, albeit oblique inquiry, and was all the response they needed to consider their assignment fulfilled. Fearing a further tongue-lashing, they made a hasty retreat.

On reporting their findings to the Vestry, and in turn the rector, who in turn reported it to the local bishop, it was determined that for the sake of the Lord and the church, our young god-wallah Nigel Wellbourne was to be relieved of his duties. In other words, he would get the sack … with prejudice, and would of course be asked to turn in his collar. His career in the ministry was soon to be over. To the congregation, it would be announced that Wellbourne would be returning to London to care for an ailing aunt.

(Oy, got the chuck, eh? Harsh, but as you've told me many times, the church invariably finds it best to avoid anything that detracts from its reputation, even if that means offering alternate facts. Isn't that so?)

[Regrettably, yes, though 'alternative facts' would be to some an oxymoron. That said, in no event do I approve of what was done to our man Wellbourne, at least in the manner with which it was done. How is it that the church oft preaches of forgiveness, but finds so little of it in its own stores?]

Following formal internal proceedings, and with much prayer, Nigel Wellbourne's employment with the Anglican church was summarily terminated. With no income, no prospects, a pregnant lover, and in the face of community outrage, Nigel was immobilized for a time. Absent recourse, and with the Lady's approval, Nigel subsequently moved in with Gemma. In a private ceremony conducted by a scorned official of the church, Nigel and Gemma were wed in the Lady's expansive garden, and despite their irregular start at marriage, fully enjoyed a glorious day, ignoring the complexities that now enfolded them.

With gracious reassurance, they resided for a time with the Lady, and Nigel began work as her gardener. In the requisite time, biologically speaking, Gemma bore a daughter, whom they named Gracie. As their familial bonds grew, the family Wellbourne enjoyed a period of contentment, residing in the sheltered luxury of a verdant English countryside.

Ere long our Nigel grew restless. Despite what he discovered were the pleasures of simple work, and with ample time to think, it had occurred to him that he was failing to make any sort of mark on his grander ambitions. It was his dream, as it had been since his days at Wycliffe Hall, to in some way influence the course of human kind.

Admittedly, these were grand ambitions, if only vaguely stated. But determined he was, and so it came to him at last that he should announce to the Lady that they were contemplating a move.

From the outset, the Lady reveled in the company of this budding young family. They had brought hopefulness and vitality to what was for her up till then a repetitiously stodgy life. While married to Lord Burgess, she had been constrained by the stultifying strictures of social convention, which left little room for personal expression, and a natural proclivity toward battiness.

(*At battiness, she seemed to later excel.*)

[**Now, now. Weren't you saying just a minute ago something about 'judge not …?' As you know, we must allow for a range of human expression — that is, within certain bounds.**]

That the Wellbournes might leave risked a life that was beginning to fall into place for the Lady. Their departure could be her undoing, so much so that she had even contemplated leaving with them. Yet her entanglements to the manor would in no small way bind her to the land. Thus she was compelled to rely on pleading, but her entreaties would in the end go unrequited. Torn between the comforts of manor living and the prospect of enriching the human tide, with Gemma's blessing, Nigel determined that they should at last make a move back to London, where his prospects for advancement would be enhanced. Nigel had it in his mind to engage with one of his former professors such that he might divine some kind of plan for gainful employment.

After they had found a suitable flat, financed by the Lady's generosity so that Gemma's small inheritance could be preserved, Nigel made contact with Professor Gantworthy, one of his former professors of mathematics. He and Nigel had formed a bond of friendship outside the classroom, owing not only to Nigel's facility with abstract mathematical concepts, but also the fervency he had shown toward aims that rose beyond those of other talented lads his age. The professor knew that Nigel was not so much interested in attaining to a comfortable station in life, but rather by perseverance and industry, he was seeking to have a grand but as yet ill-defined impact that might grace mankind. Such were matters only for the very best minds, the kind of minds that should be reserved for governing the paths of others, the professor thought.

"Nigel, how grand to see you once again, my boy. How is it with you?" The professor greeted Nigel with a fulsomeness he customarily reserved only for his closest friends and colleagues.

"It is well, it is well indeed, Professor Gantworthy. I have only lately been married, and we are fortunate to have a young child, a daughter named Gracie," he replied. Of course, the Professor could see that Nigel had an itch that would need scratching, but held off his inquiry into the matter until the social niceties had been observed for a time. "And what of your wife, pray tell? What of her background, may I be so bold as to inquire? Only the best for you, I should think!" he exclaimed with bonhomie.

"She is as a ripened fruit, I must say, a lass in full flower, exceptional in beauty and possessed of the keenest natural abilities. Her father was, though not adequately illumined, a gentleman of some rank in the Queen's Guards, who later attained to the vestry in his local parish. Having been widowed and in turn dying while still quite young, he left my Gemma an orphan in her late teens. It was not long thereafter when I met her, then an attendant to a grand Lady of Letchworth, that we were to become engaged, and thus married."

"So, then, what have you been doing with yourself?" the professor inquired.

Avoiding a direct response, "That is exactly why, aside from a long-held desire to see you once again, I have come. Between posts, I am hoping, yearning in fact, to find something suitable to both my education and ultimate aims."

Professor Gantworthy was well aware that after graduation, Nigel had pursued an education in the ministry, and correctly deduced that something had gone awry. Observing that Wellbourne was disinclined to answer his direct question, the professor judiciously chose not to pursue that line of inquiry. "And, what might those aims be?" he asked.

"Well, lately I have been reading much of the works of Galton and Pearson, particularly the latter. It is their work in pursuit of biological inheritance of, how shall I say, the more noble human attributes that interests me. I find Pearson of interest in particular, because he has moved beyond the mere counting of biological phenomena as with Galton, to the correlation of traits among the weaker strains. This to me would be of great interest. It is, as I think Pearson is known to have said, a matter of great national importance that we focus on fostering good stock among our people. Perhaps I should say it differently — we need to give more suitable strains a better chance of prevailing."

"Ah, yes, it's eugenics for you then," Professor Gantworthy summarized. "Quite so, quite so. I think this might well be a good fit for you, combining as it does your mathematical acumen with your philosophical bent. Indeed."

"Can you then be of assistance in that regard?" asked Nigel. "I thought it might be possible for you to have contacts that could guide me in that direction."

Nigel seemed overly solicitous to the professor, both in the urgency of his tone and the way he had moved to the edge of his chair as he spoke. The professor had seen this behavior in many of his students, and did not take it as anything out of the ordinary for one who had not yet discovered his position in life, but who earnestly sought a beginning.

"As a matter of fact, I might," he said. "I do know Pearson, if only professionally. We have provided his eugenics laboratory a number of students — emmm, some graduates as well. They do the bulk of his calculations, and the more mundane office work. I think you might find the statistical work somewhat beneath you, but perhaps there is something there for which you might be better suited. Anyway, I would direct you to the Galton Laboratory for National Eugenics right here at University College. It's over on Gower Street. As you may know, Karl Pearson took over the operation of the Eugenics Record Office. It had been headed by Edgar Schuster, a zoologist who was a somewhat dilatory man, and thus largely a failure at his work. Anyway, if you like, I will correspond with Pearson, and offer a letter of introduction."

"That would be most helpful. I would be forever in your debt," Nigel answered.

"No need, my good fellow. Perhaps you will at last find the beginning of a great career, one in which I expect you to make a great mark." Turning away before he concluded, he resumed shuffling through the stack of papers on his desk. "In any case," Professor Gantworthy said, "greet your family for me, and do please bring them by some time so that we might become acquainted."

"I will, sir. I will. I think you might be as enchanted with them, as I myself am," Nigel enthused.

"I am sure I will, Nigel. Oh, one last thing; I do not think it necessary for you to mention your work in the Church of England. I shan't. Take care, and good day."

"Good day, sir, and thank you," Nigel replied, donning his coat. When he shut the door to the professor's office, he immediately reflected on how it was that the professor might have learned about his work in the church. It was sufficiently painful to resurrect memories of that time, so he immediately purged the thought. Nigel had proven capable of all but erasing memories of those things that no longer fit his self-image. It was a peculiarly exceptional ability, and one that would prove useful for most of his life.

(*He had the ability to compartmentalize, I think.*)

[**Most assuredly. It's a trait of the high-minded, especially those who have something to hide. I see it most in the political class, but shall not go there just now.**]

CHAPTER II

Nigel began work at Karl Pearson's well-funded Eugenics Record Office at University College on May 1, 1910. After a brief orientation during which he was told, "The principle goal of the ERO is without political agenda, but is in fact a search for truth," his initial assignment was to perform the complex calculus linking what were thought to be inherited characteristics with the proximity of familial relationships. He worked in a room with long tables amidst mostly women. Pearson employed top female graduates in mathematics because he was able to pay them less. Although the makeup of his co-workers required some adjustment, in time Nigel adapted to the personal chatter, and was able to engage in conversation during rare interludes between silent working. His duties required intense concentration, as the computations were laborious. Distractions easily produced errors that were hard to locate after the fact. While the women worked without complaint, Wellbourne grew weary of the tedium. Fortunately, each worker was given a complete data set regarding some particular human characteristic. Nigel had been lucky that he was initially assigned calculations related to observed intelligence. He found the subject of his particular calculations to be of much greater interest than studies of the more distasteful human characteristics, such as drunkenness, disease, licentiousness, physical malformation, nomadism, and similar symptoms of wantonness and degeneracy. Such were the range of "defectives," as eugencists termed it, that were studied at the lab. To keep himself fully engaged, he looked for ways to improve the research process. Not long after beginning his job, Wellbourne observed that comparing subjective ratings of student intellect with the proximity of familial relationships would not enable the precise measurement that the nature of the problem warranted. Absent the proof of a Gaussian distribution, and with a relatively small sample of data, he repeatedly asserted to anyone who would listen that Spearman's statistical approach would have been a better choice.

[Of such finicky parsing are scholars made.]

He had said so in several formal letters to his superiors. After first being taken aback by criticism from a new employee, they ultimately conceded his point. It was a bold move, given the well-known rivalry between Karl Pearson and Charles Spearman, acknowledged leaders in the field of statistical computation.

Although the ERO's calculation methods were not modified, his superiors began to pay closer attention to Wellbourne. They had seen in him something that, though peculiar, might augur a more significant future role. Not long after he sent his last letter on calculation methods, Wellbourne penned another recommendation. Examining the files where the field-workers' case notes were kept, Nigel concluded that the coding system could be improved. He proposed that each case questionnaire would be given a five-place category code. The first digit, on a scale from one to four, would indicate intelligence, with 1 being superior and 4 deficient. The second digit would be numerically coded to indicate temperament, such as lunacy, pauperism, criminality, promiscuity, well adjusted and the like. The third code would indicate morphology, or peculiar physical characteristics, such as head or extremity deformities, or movement anomalies. The fourth would indicate any special ability, such as math skills, musicality, or athleticism. The fifth would be a familial or surname indicator to indicate family groupings. Admittedly more laborious, his approach would enable researchers to quickly spot patterns within and between characteristics. One could see at a glance at the files that those with exceptional math skills, for example, tended to have higher ratings of intellect.

Wellbourne's superiors were so impressed with his thinking that they hired additional workers to put his coding system into effect. He was placed in charge of training his co-worers, and immediately hung a large sign that explained his coding structure over the work area.

Before the year was up, Wellbourne, who had also begun to travel extensively throughout England to document a wide range of inherited degeneracy, was recognized for his superior contributions. As a result, he came to the attention of Karl Pearson. Pearson had one of his assistants invite Wellbourne to a speech he was to give at the Royal Society. Wellbourne valued the invitation for the direct exposure it provided to the leaders of the eugenics movement. On the cusp of a new decade, he could see that the movement was to take on a much more prominent global position among the intellectual elites.

The night of Pearson's lecture, Wellbourne was detained by his young daughter's illness. Concerned for the health of two-year-old Gracie, Gemma requested that Nigel abandon his plan to attend. Her insistence added considerable tension to his evening, but once the situation at home was seemingly under control, he departed in haste, and arrived at the lecture hall well into Pearson's talk.

"Ladies and gentlemen, as you may know, many acclaimed figures such as G.B. Shaw and the noted Havelock Ellis, as well as your humble speaker, can assert with confidence that class distinctions are indeed barriers to optimal marriage. It must be that our values, our social structures, and indeed our government should be constrained in ways to limit such class distinctions. Accordingly, it is only through optimal marriage between the classes that we can begin to achieve our ultimate aim of scientific propagation. Let me just add parenthetically that my dear cousin Darwin, a social Darwinist himself," he said chuckling, " eventually came to believe that biology is destiny, and in that he, too, like many here present, should have by now much more unreservedly concluded that genius is the greatest of all biological inheritances."

Thunderous clapping punctuated by boisterous shouts of approbation erupted from the gentlemen there assembled, after which Pearson took a few questions.

"Sir, allow me to express my appreciation for your talk. Now, if I may, where do you think Great Britain stands in all this? In other words, what must our nation do?"

"Thank you for those kind words. Let me just say that the ascendancy of the nation, the ascendency of man himself, depends on morality, and morality in turn depends on biological fitness, which I believe can only be achieved through national socialism. If I may be so bold, let me just say that thus far our present civilization has only made the world safe for stupidity."

(*I gather that by saying this, he is alluding to social welfare programs.*)

[**Shush!**]

Shocked by the assertion, which in its abruptness pointed to a profound truth as they saw it, after a sustained silence, the audience guffawed loudly. It was as if they had finally gotten a most clever joke.

Pearson went on, "Can we not say, then, as in Mendelian science, that just as we have improved plant and animal species, so too we can improve the race of men? But that work must begin at once, and must not stall because of the faint-hearted interventions of those who would maintain the status quo. For even now, my research shows a nervous weakness among our people, a neuropathic taint, the balm for which can only be an improved germ plasm. The revised laws of heredity suggest that human populations can be permanently improved only by biological manipulation. As we have shown, the force of heredity appears to be powerful enough for features like intelligence, so as to dictate selective breeding as the only means of achieving greater social strength. And in consequence, the nation must intercede, because Britain has ceased to breed for intelligence. To that end, it must be social imperialism for our nation. Even now, the demographic trend is dangerous. Population growth comes from the least fit segment, the unfit, which is in my estimation a product of capitalism, and its con-comitant demand for cheap labor. Schools cannot correct the problem. No amount of education will help. You must breed it. "

More cheers, as our man Wellbourne was similarly swept up in thrall of Pearson's stunning genius. To Wellbourne, Pearson was a first-rate man, fundamentally sound, and among the ablest of fellows. In thinking that, Wellbourne had at last made the connection between his own vague thoughts formed so long ago, and this new mandate to elevate man himself beyond mere abstractions, to a more pure and noble self. Eugenics, Wellbourne confirmed then, must indeed be his life's purpose, his mission, and his work. He could not wait to get back to the office, but first he must go home to his wife and ailing daughter.

(Bollocks! That was just crazy talk from Pearson. His smug certitude was simply breathtaking. Hubris such as this ought to be rebuked with prejudice, if not met with some judiciously applied discipline, perhaps even a personal calamity. With such contempt he regards natural human variation!)

[Now, now. As man is born in liberty, he must be permitted error and chance, thereby revelation, and finally creativity and change. Perhaps then he may find redemption. I have always thought it best that under such circumstances events should just play out. Calamity can be permitted at an appropriate time.]

(*Forgive me for saying so, but there have been times you seemed to relish calamity. Isn't now such a time? Too often of late, it seems to be patience for you.*)

[You judge harshly. Calamity, though seemingly random, is a device I allow for some greater purpose, though I confess to enjoying a good dust-up from time to time. Anyway, as I say, it is often wise to let situations develop.]

By the time Wellbourne had returned to his flat, his fervor had cooled somewhat, but he was nonetheless inclined to share with Gemma his reinvigorated enthusiasm for eugenics. However, seeing her holding a damp cloth over the forehead of ailing Gracie, his passion subsided, and he inquired with gravest concern about his young daughter.

"Her fever has broken, I think," Gemma offered, noticing that Nigel was staring at her hand on Gracie's forehead. "I'm just refreshing her, poor dear. She had been somewhat agitated, but has settled back to sleep. I think she will be fine by morning. I suspect it is just a normal childhood fever."

Relieved, Nigel kissed Gemma's hand. "You are such a wonderful mother. I count myself blessed for having found you."

"How was your meeting?" Gemma inquired, rising to her feet.

"It was ... it was inspiring," he said, emphasizing the last word. "I will tell you more tomorrow. For now, I think I should just retire. I am very tired. I have a big day ahead. Still, Gemma, I can scarcely wait to share with you what I have learned."

"Yes," she said. "You do look tired. We can talk tomorrow, then. Good night, dear."

Nigel rose early the next morning and hurried to the office. His steps were light as he eagerly anticipated the day with a renewed enthusiasm for the mission. When he arrived, there was a hubbub just inside the front door.

"Oh, Davenport," Pearson said, "do allow me to introduce one of my most promising associates. Wellbourne, this is Charles Davenport, one of the movement's leading young tigers. Davenport here," Pearson said to Nigel while gesturing with his thumb toward his guest, "is over in America building on our work. Charles, this is Nigel Wellbourne. Wellbourne is the one who has helped us make substantial progress in our research through a rather innovative coding system. Needless to say, he also excels at computational work."

Standing more erect, Wellbourne addressed the man for whom he had already acquired great respect. He was well aware of Davenport's advances in the field, and had read Davenport's much-hailed book, Heredity in Relation to Eugenics, which had been published earlier that year. "I am pleased to make your acquaintance, Mr. Davenport. We have heard so much of your work."

"Oh, do please call me Charles, Mr. Wellbourne. Say, before you arrived, Pearson here had been telling me about your new system. It seems quite a good advance. It should make quite apparent, even in the absence of calculation, the linkage between family members' inherited traits. If you ever come to the States, perhaps you should consider coming to work for us!" With that, Davenport guffawed uproariously,

Pearson laughed, and those in the immediate vicinity chuckled along with them, as Nigel blushed.

Jealous of the attention Wellbourne had received that morning, his male coworkers sought to burst his bubble by sharing somewhat disparaging gossip of their employer, someone whom Wellbourne had quite obviously begun to adulate. "Say, Wellbourne," one of the few men in his work area said, "I don't suppose you've heard that our fearless leader once formed a Men's and Women's Club here in London. Seems he wasn't getting enough on the side, so that under the guise of fostering intellectual discussions between the sexes, he sought physical encounters with some of society's leading lights. Cleverly, he accomplished this by holding small, private discussions about such things as sexual relations, contraception, free love, self-gratification and the like. Quite titillating, I should think."

Nigel did not have a response. It was simply too unbelievable. Pearson was too much a religiously rigid figure to have been involved in such a debasing endeavor. In a sign of contempt for their loose talk, Nigel abruptly turned away, and resumed his work.

In the weeks that followed, having completed computations on data relating to the correlation of intellect with the degree of family relationships, Wellbourne was asked to submit an abstract of his work to the lab's publishing group. Because those who worked on calculations were as a matter of course separated from those who published their findings in Galton and Pearson's *Biometrika*, this was a great honor for Nigel. He plunged vigorously into the details of analysis, and labored exhaustively to write the brief summary of his findings. He had spent sleepless nights in the lab completing his work, much to the chagrin of Gemma. When he returned home, spent as he was, Gemma was outwardly supportive, but internally overcome with worry for his health and mental state. After a suitable interval to recover, Nigel was at last enthusiastic in sharing with Gemma his conclusions, and allowed himself the privilege of boasting about his contribution to the journal article. Gemma, of course, gave due recognition to his effort, and was in every way approving, despite the burden that his work had placed on their home life.

When the issue of *Biometrika* had been produced, it was with profound anger that Wellbourne realized that no attribution for his contribution had been provided. The name of one of Pearson's oldest associates was listed as sole author to the work. It was for Wellbourne an unpardonable slight. Once again, he contemplated a career move. When he returned to his flat, his very countenance betrayed a grievous wound. Gemma, of course, could see a storm brewing, and immediately sought to calm the turbulence that would ensue.

"Dear," she said, "I can see that you are very tired. Please sit, and allow me to brew a fresh pot for us, won't you?"

Despite his agitation, Gemma's soothing words had a moderately calming effect on Nigel. After a due course of mumbling perturbation, followed by several sips of tea, he was finally able to calm himself enough to converse.

"Tell me, dear, what manner of discord has cast its shadow over my beloved husband?" she asked. "You are too prized for your talents for there ever to be anything that might distract you from your worthy endeavors."

"Oh, Gemma" he said, " This is all so painful for me. I've worked diligently for many months on a project of great importance. I've traveled throughout England to institutions for the feebleminded, trying to find among the detritus of human relations some hereditary linkage. I've walked the wards of Earlswood and West Riding, ridden the train to Miles End, and trudged the farm fields at Leavesden, interviewing, documenting, and calculating the twisted branches of hereditary failure. I have labored over blindingly tedious calculations, invented and implemented a new coding system — all this without complaint. And in my hour of greatest labours, I am asked to contribute to a leading biometric publication. Yet, is there then any recognition? None, I say none! It is disheartening. Troubling, indeed. I should think that some mention, however meager, might have been made, but no — I'm relegated to a non-being, while someone barely my senior is given sole credit. A veritable lab serf is what I've become. 'Tis a travesty, Gemma, a travesty, and I shall not accept it!"

(*'Feebleminded'… 'human detritus'… 'twisted branches of hereditary failure.' Aren't you in the least insulted by such talk?*)

[**Pity our poor Wellbourne. Troubled, yes … I have been for some time. But he used common vernacular, and his motive, I think, has always been to improve the lot of mankind — a worthy goal in and of itself. That is not all bad; though, I confess it might be well for you to have counseled him a bit.**]

(*You seem unsure of his motive. That is not like you. In any event, I have been whispering in his ear for some time. Listening was never his strength. He is always too deeply absorbed in his own thoughts.*)

[**Enough now. Please allow me to continue.**]

"Oh, Nigel," Gemma said as she came to his side and rested her hand on his shoulder. "It must have been some oversight. There cannot be another explanation. Surely they know your contributions. Talk to them. I know they will assure you of your value."

"I will talk to them, all right. I will give them my notice," Nigel said. He rose, and began pacing, leaving Gemma's hand hanging loosely by her side.

(*This was all such a great calamity for poor Nigel. Where was your restraint this time?*)

[**Impertinence does not become you, Counselor. Again, I advise patience. Things work out.**]

CHAPTER III

THE NEXT DAY NIGEL AROSE REFRESHED.

(*Not just refreshed, but more tolerant. One might think he had been visited in the night by some wise spirit.*)

[You give yourself too much credit.]

Nigel apologized to Gemma for his impetuous outburst the night before. He grudgingly acknowledged that perhaps he had been wrong in his assumptions about Pearson, and admitted that it would be better to discuss with Pearson how he might be given the opportunity to write his own research paper.

To Nigel's astonishment, Pearson was receptive to Nigel's proposal, and went on to compliment Nigel's work. He suggested that Nigel present findings of a recent research project at an upcoming meeting of the Eugenics Education Society. Days after the event, Pearson surprised Nigel by publicly congratulating him, and said that his talk had been especially well received by the Society. Recognition was all Nigel needed to restore his enthusiasm for the work.

Several weeks after the presentation, Pearson called Nigel into his office. "Nigel, my boy, I have a new assignment for you. It's one of great importance, and I think you will find it most intellectually engaging. If you agree, I would like you to analyze some data, which we have been accumulating for the last two years. When you are confident in your conclusions, I would like you to submit an article to *Biometrika*

… under your own name, of course. I hope that meets with your approval, sir," Pearson said.

"And what, may I ask, is the subject matter for this inquiry?" Nigel replied.

"Pigmentation in Relation to Selection and Anthroprometric Characters," Pearson answered. "It is critical that we gain an understanding of, how shall I say this, anthropological differences and their effects, if you take my meaning. I think you will find this work quite stimulating, to which I might add, we shall no doubt find your conclusions most illuminating." And so it was that Wellbourne labored for the better part of a year to verify that there were indeed differences between the races, skin color being foremost among them. The data confirmed what the real experts already knew to be true.

While the esteem in which Wellbourne was held grew markedly, the nature of his work soon took a decidedly different turn. Contrary to his preference for studying hereditary causes of intellectual capacity, particularly by taking advantage of Binet's latest advances in the emerging field of IQ measurement, he was confined to studies relating to a host of family diseases. Science had progressed well beyond morphology, mere descriptions of physical properties, and moved past Galton's metric approach to science, in which living things were sorted, counted and measured. By then science had begun to determine through correlational studies the degree to which characteristics related to each other. With their newfound mathematical tools, scientists set out to correlate nearly everything in the natural world, even phenomena which to a layperson might have had only spurious connections. It was through such zealous work, for example, that poverty was determined to be the proximate cause of tuberculosis.

(*Ahem. And the moon causes lunacy!*)

[I must say, you show a high degree of cynicism about scientific progress.]

(*Progress, you say!*)

[Look, these are neither gods nor automatons we are dealing with. Time must be permitted for discovery, else why would we need more scientists?]

"Nigel, love. You seem restless once again," Gemma observed. "Have you troubles?"

"Oh, Gemma, I have grown so weary once again these last months. My work has once again grown unfulfilling. I do not know how much longer I can delve into physical affliction and degeneracy. Is it not much more important that we focus on the mind, that which governs all progress, the veritable seat of all humanity? Only then can we begin to address the afflictions which plague our species."

"Are you then referring to your studies of intelligence?" Gemma inquired, knowing full well that had always been Nigel's preoccupation.

"Yes, of course, my love." he said. "Even now, Binet has made great progress in measuring intelligence. We ought to use his work. But Pearson has been insistent. He says that the subject has been thoroughly covered. Oh, I have grown so despondent of ever altering his thinking. Perhaps it is time to take Davenport up on his offer."

"His offer?" Gemma asked.

"Yes, you'll remember we had a chance meeting at the lab nearly one year ago. After Pearson introduced me, Davenport suggested that if I should chance to make a move to America, I should see him about a position. Cold Spring Harbor, I think he said, is where his lab is located."

"Where is that?" Gemma asked, now distracted by Gracie who, too young to have started school, was already working on a problem of long division. She wanted her mother to check her solution.

Before Nigel could answer, Gracie turned her face up to her mother and asked, "Mummy, did I do it correctly?"

"Why yes, you did, darling," Gemma answered.

"Daddy, do you think I'm smart?" Gracie asked.

"Yes, dear, just like your mum," Nigel answered.

"And just like you, Father," she replied.

"Long Island. Near New York City," Nigel blurted in response to Gemma's earlier question.

"What?" Gemma asked him. "To what are you referring? Oh, yes, the Cold Spring Harbor laboratory. Are you even certain that you can have a job there? Might this have just been polite conversation?"

"Certainly there is position there for me. Davenport said so himself. Don't you think the opportunity might be grand?" Nigel continued. "A change of venue might do us all good. T'would be much less sooty," Nigel intoned, hoping to appeal to Gemma's sense that London wasn't the healthiest place to raise a family.

"Well," she said, "you know how I feel about coal dust. It is not beneficial to our health, particularly with our young ones. It seems to cover everything in our little home."

"Our young ones?" Nigel asked. "To whom are you referring?" he continued, for he had recently suspected that their family circumstance might be changing.

"I imagine you already can guess, Nigel," Gemma said. "It is true, I am — we are expecting," she added, tenderly placing her hand on his. "Are you delighted?" she asked hesitatingly.

"I am, dear. I most decidedly am," he enthused. "Despite what might turn out to be a challenging adventure, what with a move, finances and all, a second child will still be most welcome." He said this for Gemma's benefit mostly, as he expected that the demands of a new position would afford him little time at home. Although Gracie had required so little of his time, he could not be assured that a second child would not be a burden on the whole family. But, for the sake of Gemma's emotional peace, he thought it best at this time to offer a statement of support, even if it was offered with a cautionary note.

The following week, as Gemma busied herself preparing a morning meal for the three of them, and while Gracie was not yet awake, Nigel girded himself for what was to be a most delicate conversation.

"Gemma," he said, "you do know I love our little family, don't you?"

She stiffened as she continued to roll out dough for fresh scones. "Of course, Nigel," she said. "Why would you ask such a question, even before your morning tea?" Her faint attempt at levity was missed by Nigel. Preoccupied with the terrible thoughts he was about to share, he did not hear her response.

"Gemma," he hesitated before continuing, "there is something I must ask. It is most difficult for me … for us."

Gemma turned abruptly, not knowing what he might say next. "What is it, dear?" she implored, trying her best to mask her fear with a sympathetic tone.

"You know how important a move to America might be for my career — I should say, for our future."

"Yes, of course I do. I know it means a lot to you. You have so much to offer," she assured him.

"It's just that finances are tight. I fear we shan't have enough to finance the trip, especially with a new child on the way." His voice began to tremble.

"We can use my inheritance, if that is what is required," she said. Her voice took on a more business-like tone. "We can make do. Anyway, I know we are both excited at the prospect of a new baby. Gracie has added so much to our lives. She is such a grand child, and so smart. Another child will be double the blessing." Neither Gemma nor Nigel saw Gracie, who by then had silently appeared in the shadow of the doorway to the kitchen.

"Gemma," Nigel said, finally screwing up the courage. " I fear I must just speak my mind."

Gemma furrowed her brow, and said, "Go ahead then, dear."

"Gemma, I'm wondering if we shouldn't consider delaying a child for a time."

The intimation was not lost on Gemma, who recoiled at the mention. Leaning back against a counter near the stove she asked, " Delay? Just how might that be accomplished? You mean an abortion? Isn't that what you mean?" Her voice rose in anger as she spoke, and in mere seconds tears filled her eyes. "How could you even think that?"

"It's just that, assuming quickening is a ways off, we might consider ahh, em — you know, Beecham's Pills, or some such. Lydia Pinkham's Vegetable Compound if you prefer. I know this is delicate, but people do it all the time. This would only be a temporary measure. Once settled, we can have another child."

(*Lydia Pinkham's Vegetable Compound! Vegetables? Do you not think it amazing that people are so easily duped by innocuous sounding names and deceptive pairing of the bad with the good?*)

["They know nothing, they understand nothing; their eyes are plastered over so they cannot see, and their minds closed so they cannot understand."]

Gracie lingered in the shadows, heard these things, and stored them up in her heart. Even at four, she understood something horrid was being discussed, and like her mother, was frightened. The memory of their conversation would remain with Gracie for the rest of her days.

"Nigel," Gemma said, finally asserting herself. "I will not hear of it. Temporary measure, indeed! Not to the child. The thought is ghastly, and ought as well be to you." She was all but yelling, and Gracie took the change in tone to be an opportunity to slip back to the bedroom. After a time, sensing that the argument was finally over, she left her room. She closed the bedroom door loudly to signal her emergence.

"Gracie, dear," Gemma said, "good morning. How did you sleep? Your father and I were just discussing our excitement at the prospect of a new baby."

"Yes, mummy." Trying to overcome her fear, and to show enthusiasm, she paused for the briefest moment, then finally said, "It is a matter of some excitement. I can tell." She looked up knowingly at both her parents with only the slightest hint of a smile, then remained silent. Her parents could see that Gracie had overheard them. They abruptly ended the conversation.

In marriage, even the smallest comment or gesture offered in the heat of passion can have lingering effects: from an emotional cut, even an abrasion, a deep wound can develop. Some heal, but leave scars. Though they may be covered over, wounds of unresolved conflict never heal.

No more was ever spoken of the matter. Nigel quietly went about his work at the lab, as he and Gemma laid up every spare farthing in order to accumulate enough savings to begin a new life in America.

For Nigel, the prospect of another child at this time was a complication, but it was one that with determination could be mitigated. He could endure inconvenience so long as they were working together on the goal of moving to America. And though he knew Gemma was hurt by the callousness of his suggestion, he was sure she would get over it soon enough, for maternal instinct almost always covers over all manner of sin. He was certain of it.

Gemma, on the other hand, attended not only to new life, but also a grudge that she knew she would have to endure, if not bury completely. An abortion was never something she had considered. It wasn't that the idea was foreign to her. She had heard too many stories to be so naïve. It's just that in her mind a baby, any baby was too precious a gift to ever consider terminating. To add to the psychic wound which Nigel had inflicted, Gemma was hurt beyond measure that Gracie had heard their conversation. Gemma was sure that even if Gracie did not know the purpose of abortifacients, whose purpose was masked by an innocent sounding name, she would certainly have understood that her parents were quarrelling about something that had ominous consequences. Gemma could not recall an occasion when Gracie might have seen her parents disagree, let alone a time when her mother cried in response to something her father said. That Gracie had so skillfully tried to mask her fears when she appeared in the kitchen was all the more troubling. Gemma knew that Gracie would carry the fear deep inside. Yet it was something that Gemma would never be comfortable discussing with her child.

Married life for Nigel and Gemma soon reverted to its normal course. Rebuked by Gemma for his heinous plan, Nigel approached his work with compensating vigor, and for a time without complaint. He continued to travel throughout England collecting endless family histories of the enfeebled and the afflicted. In that time, he encountered more of what he termed 'human wreckage and social dysfunction' than he had ever thought possible. To him, this blessed plot, this earth, this realm, this England was beginning to look less like a well-groomed garden than an untended back-alley filled with rubbish and thorny undergrowth. Inspired to make a better world, yet forlorn over its present state, he became all the more committed to finding means to overcome the want and neglect that had so imprisoned the poor and hurting of England. It was obvious to him, as it was to so many, that government redistribution of wealth did not seem to be doing the trick. And though he and Gemma spoke little of their impending move to America, both made the necessary sacrifices to make ready for the trip.

For her part, Gemma acted as if nothing had ever been offered but expressions of joy and anticipation at the forthcoming birth of another child. She busied herself accumulating the necessities that a newborn might require, and managed the household in Nigel's frequent absences. She considered it her duty to be pleasant and supportive to Nigel on his return, and to be in every way agreeable to his predilections.

For Gracie, it was a time of maximum bonding with her beloved mother. It was Gemma who had served as the dominant parent both in terms of social and academic nurturance. Born of superlative native intellect, Gracie was already an apt student. Invariably she was, to Gemma's friends and acquaintances alike, at once astonishing and charming, owing to her prodigious talent. All agreed that it was remarkable in one so young. To be certain, there was stress on the Wellbourne family, but it lay below the surface. To the outside world, all was well, and their lives were abundant with promise.

"Nigel," Gemma began in a conversation some months after his unpleasant suggestion, "do you not think we have been so very blessed by our child? She is such a willing student, and readily grasps everything I can teach her. Perhaps we should consider giving her a private tutor."

"Most certainly, Gemma," Nigel replied. Immediately Gemma grew expectant, but her hope was dashed as Nigel continued. "Gracie is a blessing, and she does profit well from your teaching. I can think of no one better to continue her lessons. As for a private tutor, I think it best for us to hold off for a time, at least until after she begins school in America. Then we can better judge the need."

"Of course, you are right. I just wanted to convey my sense of gratitude at the blessing of such a gifted child. She requires so little discipline. I know you agree," asserted Gemma.

"I do, Gemma, I do. We should count ourselves fortunate indeed. Let me hasten to add, as our work in the lab has so clearly shown, she is the product of good breeding. Which breeding, I think, arguably goes back generations on both our sides. Good protoplasm, he enthused. "That's the ticket!"

Gemma was not startled by this punctuating assertion. She had heard it before. She knew he was merely repeating something that was often said at his work, something no doubt offered as a cheer to motivate the workers. Good protoplasm, indeed.

"Nigel?" she began after a quiet interlude in which they exchanged loving glances. "I am so looking forward to our new baby."

"I am as well," he said, looking over the top of his book.

"We have every reason to believe that he or she will be just as lovely a person as our Gracie, and just as gifted besides, don't we?" Gemma asked.

"Yes, dear," he concluded, "we can be certain of it." They once again looked deeply into each other's eyes. He reached over and squeezed Gemma's hand lovingly.

Like all expectant mothers, Gemma was hopeful, and filled with love at the prospect of emergent life, but at the same time little fears crept in — a stillborn child, birth defects, a persistently weak constitution perhaps. In moments of weakness she might fret only briefly at the possibility that something could go wrong. Nine months was a long time to live in expectancy. Still, she was able to manage her fears. Raising Gracie and taking care of Nigel invariably blocked anxieties that only too briefly emerged. Of course, she would never share her anxiety with anyone else.

Seven months after their most unpleasant conversation, Nigel, Gemma and Gracie were at last prepared to embark on their new adventure. With Gemma's inheritance and the money they were able to save, they would have enough to purchase tickets for the voyage and set up a small apartment in America. Their lives had progressed smoothly, with no troubles appearing on the horizon, save for Gemma's growing fatigue in the month before departure. Both they and the doctor attributed this to the stress of the move and the normal course of pregnancy. There had been a minor disagreement about their, or I should say Nigel's, plans for shipboard accommodations. Nigel had, without the benefit of speaking first with Gemma, assumed that they would be traveling with second-class passengers. Needless to say, first-class opulence was out of the question. It was only natural in his mind that mixing with the better classes and not necessarily the very rich was most befitting his assumed station in life. Then, of course, there was the matter of traveling with a small child. It just would not do that Gracie should be exposed at such a tender age to, how shall it be said, elements that were not suitable for one so young. But Gemma knew the cost of the tickets meant that on the other end of the trip they would be constrained to find an apartment that would cause them to mix with "the elements" which Nigel had sought to avoid. Moreover, Gemma would be limited to a household budget that would challenge her ability to provide reasonable comfort and nourishment, let alone avoid the pestilence that was no doubt prevalent in such abodes. Said another way, she feared that second-class tickets might all but deplete her inheritance and their savings, and would possibly risk in the end a life on the streets. His ill-considered decision had caused another rift in what was an altogether temperate relationship. This time Gemma found her voice, and Nigel was forced to concede his irreconcilable mistake.

Nigel's resignation was graciously received by Pearson. He not only acknowledged Wellbourne's contribution to the lab, but also assured Nigel that he would forward an enthusiastic recommendation to Davenport. The good news of this, once shared with Gemma, relieved some of the tension which owed to his previous financial blunder. Perhaps, as he offered to Gemma in verbal recompense, he might possibly be eligible to receive an advance from his new employer.

When at last the day of their departure arrived, the family Wellbourne was in a jolly mood. They were brightened by the prospect of beginning anew in that dawning time between fading past and emergent future. The warm greetings from the shipboard staff, and the buoyant mood among the second-class passengers, only served to elevate their good feelings. Gemma's quite visible pregnancy brought gestures of gracious accommodation by many. Such gestures were typically reserved for only the most notable or obviously well-to-do. Whatever fatigue she may have felt before departure was easily ignored.

CHAPTER IV

AFTER A SHORT SEARCH, during which Gracie skipped alongside, Nigel and Gemma located their berth. Appraising each of the room's modest amenities absorbed their attention during the brief wait for their embarrassingly meager luggage. Two small, skirted wooden beds with spindled side rails were placed against opposite walls. Gemma ran her hand over the white linens and thin woolen blankets that covered each bed and pronounced them good. The mattresses would need to be tested, which was a job well suited to a child too long confined on the train ride to their point of embarkation. With encouragement, Gracie removed her shoes and bounced on the beds with vigor, and pronounced them good. As there was ample room in the area separating the two beds, it would naturally fall to Gemma to render a verdict on the sufficiency of the floor space. Kicking off her shoes, now too tight from swollen feet, she did a fetching fox trot, finished with an elegant twirl, and ruled that the space would be more than adequate. Dizzied by her graceful spin, she readily accepted the suggestion to rest on the chair next to the mirrored washstand. A third bed was hung from the ceiling. Nigel gallantly offered to take it, though quite obviously there was no suitable alternative. He declined their playful suggestion to climb up and test it, because it was, as he stated emphatically, unseemly for him to do so fully clothed. Sensing their disappointment in his unwillingness to play along, he reminded them that he would be testing it at a more appropriate time, despite his assertion about the danger posed by such a lofty perch. Getting no response to what he assumed would be seen as a noble gesture, Nigel was momentarily wounded. Expecting affirmation, and hearing none, Nigel nonetheless managed to pull himself together. Redirecting the ladies' attention he commented that he was surprised, and not altogether displeased, that everyone had been afforded not one, but two pillows. Rather than continue to seek a response to his gallantry, which in any event would not come, he grabbed a pillow and playfully swung it at Gracie. Gemma, who had at first opposed spending the extra money on second-class service, conceded that the extra pillows did seem to make the extra expense worthwhile. At least that is the way Nigel heard it. It is certain, though, that in the expression Gemma's tongue was firmly planted against her cheek. Never to be left out of the fun, she too grabbed a pillow and bopped Gracie who, while bouncing on a bed, quickly joined the good-natured family fray.

Once the small wooden washstand and chair had been officially approved, the only window to the outside deck had been opened and closed several times, and the overhead fan tested, Nigel suggested that the two ladies settle in. He would take a quick turn about the deck to locate the ship's amenities and report back. With barely restrained laughter, Gemma and Gracie saluted, "Aye aye, Captain."

Donning coat and hat, Wellbourne sauntered to the deck area, retaining for a time the jovial mood engendered by family play. Walking along the ship's rail, he nodded courteously to fellow members of second class similarly exploring their home for the next week. He mostly encountered men, though occasional bands of youngsters scrabbled about. It was obvious to him that his fellow passengers were not only similarly dressed to those with whom he most often associated at home, but also carried themselves in accordance with the customs of their respective station in life. That is to say, they walked in haste, and had little time for conversation. He could see they were not so different from himself and likely, he assumed, had similar breeding.

The main lounge, located toward the bow, was furnished comfortably with armed wooden chairs gathered around small tables. Divans were scattered around the perimeter. The adjacent dining room was arranged with parallel rows of tables the length of the room. Small armless chairs extended in close proximity along both sides. The tables were covered in inexpensive linens. Place settings were simple, unadorned and adequate for the task at hand. Separate men's and ladies' lounges were located nearby.

Curiosity having got the better of him, Wellbourne also explored first-class, up one flight of stairs. A covered promenade ran along the length of the ship. There, handsome grandees wearing Chesterfield coats walked arm in arm with fashionably tailored matrons of conspicuous wealth. Whereas his dining room was functional and modestly embellished, the first-class dining room was utterly grand. Around each table were from two to eight leather-armed chairs. Fine china and assorted crystal glasses were featured in an eight-piece place setting appropriate for every conceivable food item. The coffered ceiling of silvered glass reflected back to the diners their own exquisite taste.

Wellbourne went on to the first-class lounge. He found still more opulence. A glass-domed, sky-lit ceiling spanned nearly the length of the richly paneled room, around which was a colonnaded perimeter. Throughout the middle section were small table groupings. Each brocaded chair and settee was finished in the finest material, beckoning touch. Clusters of refined ladies and gentlemen discoursed animatedly of travel, politics and finance, in contrast to his own peers, who talked mostly of weather and the duration of the journey. Pausing to admire the features of the room, he had not noticed a rapidly approaching steward, who inquired as to the manner in which Nigel might be assisted. It was the curious phrasing — 'assisted' rather than 'served' -- that had aroused Nigel's defensiveness. In the instant before he answered, Nigel wondered how it was that the steward came to suspect he did not belong there. Was it his attire? Had his general demeanor tipped off the steward? His shoes perhaps were not sufficiently blacked, he thought. As if in self-defense, he thought that he was at least the intellectual equal of everyone there — and more likely superior to all but a few of his fellow passengers in second class. Nigel at last replied to the steward, "I do not require anything just now, thank you." Bowing to discretion, the steward turned abruptly and returned to his station. Fearing that he had been found out, Nigel exited the room so as to avoid an embarrassing scene.

His curiosity not yet sated, Nigel took one small peek into steerage, down two flights of steps. He looked into a large day room fitted only with wooden benches and chairs. There he observed Europeans: Southern Europeans mostly, and some Jews, with the occasional Irishman thrown in. Of course, there were many more children than he had seen on the decks above. None of these passengers were so richly attired as those above. The babble of languages and general din made conversations indecipherable. Thus informed of the ship's layout, Nigel returned to his cabin, where he found Gemma beginning to nod off. Gracie was beside her reading a book.

"Why don't I take Gracie to get some fresh air? You can rest for a bit. Supper is not for another three hours. If you like, I can order tea to be brought round at 4 o'clock," Nigel offered.

"Thank you, darling. That would be lovely. I fear the fresh sea air has made me quite exhausted. Then again, it might have been the pillow fight," Gemma quipped. Gracie leaned over and planted a gentle kiss on her temple. "Thank you, dear. Be sure to take a wrap. It will be getting quite cool soon." Gemma rolled over and was beginning to breathe heavily before Nigel and Gracie left the room.

"Perhaps you will want to bring your book with you, Gracie," Nigel whispered before closing the door. Gracie went back in and grabbed her book, then said, "Thank you, Father. You know me well," to which they both had a quiet laugh.

After a nice stroll around the open deck, during which Gracie chatted endlessly of fish and fowl and monstrous sea creatures, they adjourned to the second-class lounge for warmth and refreshment. There were now quite a few more passengers scattered around the room than he had seen on his first tour of the ship. Nigel and Gracie chose a love seat on the outer perimeter of the lounge near an older man who sat alone reading a newspaper. At first they did not notice that he wore a clerical collar. Once seated comfortably, Nigel ordered tea. The gentleman nodded to them both.

"Lovely child," the man remarked pleasantly. "Your daughter?"

"Oh, thank you," Nigel replied. "Yes, she is."

"I assumed so," the gentleman said. "And how old are you, young lady?" he asked.

Looking up from her book, Gracie replied, "I am four, but I will be five in a few short months."

"Are you excited about the trip?" he asked.

"Most certainly!" she responded. "We are going to America!"

"Indeed you are. I am as well. Whether they know it or not, everyone on board is," the kindly gentleman added. "It should be a lovely trip. The weather is expected to be decent." The man returned to his paper.

Nigel had instantly grown comfortable with the man, noted his American accent, and took the opportunity to inquire about economic conditions in America. A discussion ensued about rapid industrialization and the like, all of which was tedious and of little interest to Gracie, who by then had nestled against the side of her father, legs drawn up, totally absorbed in her book.

"I can see that she is a gifted child … so poised. Her book seems … well … advanced for one so young. She obviously comes from good stock," the reverend said. "Sir, what line of work are you in, if I may ask? What is it that brings you to America?" The reverend spoke with what Nigel presumed was typical, and somewhat jarring, American forwardness. This was clearly an opening for Nigel to talk about his work. He did not ever have to wait long for the door to swing wide.

"It is interesting that you ask, Reverend. My family and I are moving to Cold Spring Harbor, New York, where I will be engaged in research on inherited human traits."

"You don't say," the reverend replied. "Pardon my manners, sir. My name is Robert McDougal, Reverend Robert McDougal. I guess I don't need to explain to you my work, but I am interested in hearing about yours."

"Sorry. My name is Nigel Wellbourne, and this is Gracie," he said patting her gently on the head. As I say, I will be doing work on inherited traits. It is what I have been doing the last few years at the Eugenics Record Office, which is affiliated with University College - London," Nigel said.

"Eugenics, then. I've heard of it, but am not exactly sure what it is all about. Can you tell me?" Reverend McDougal asked.

"Certainly," Nigel replied. "It relates to the study of inherited characteristics, with the intent of fostering improvement in important human traits through better ... how shall I say? ... through the encouragement of more conception by those possessing distinct advantages, as opposed to those of inferior stock."

"I don't quite follow," said McDougal, who masked his astonishment at Wellbourne's words.

Nigel resettled himself on the loveseat and turned to directly face McDougal. This necessitated rearranging Gracie, who was still too deeply absorbed in her book to do anything but utter a sigh. "In our work," Nigel went on, though in a more subdued voice, "we have verified what most have generally observed. That is, the better off in society, those with more education, and of course intellectual advantages, tend to have fewer children. In contrast, those among whom we find the most problems adapting to modern society have the greatest number of children. Consequently, the problems of the latter group — alcoholism, pauperism, licentiousness, disease, for example — are passed down from generation to generation. It is a consequence of weakened germ plasm."

"Fascinating," added McDougal, whose rapt attention had caused him to put down the paper he had been reading, and turn more directly to face Wellbourne. "Do go on," he said.

"Just as we have improved agricultural productivity through selective breeding, we believe that the same can be accomplished in humans. Let me explain."

"By all means," replied McDougal. "I've never heard of this."

"Oh, yes," Nigel enthused. "Sir Francis Galton, Karl Pearson, even Darwin and many other leading thinkers have established hereditary links between the better human qualities. Such things as intelligence, moral living, even success in life have been shown to pass down from generation to generation. All this is related to germ plasm, the essential element that is in our cells. It is what conveys the traits that make us human. As I was saying, we know how to breed cattle to yield more meat or milk for a given quantity of feed. Similarly, we know that through careful seed selection we can improve crop yields. So, too, we can improve such things as human intelligence. It is already a matter of settled science."

"Settled science, you say." McDougal was taken aback by these radical thoughts, and could only now frown and offer a guttural response. Nigel turned to pat Gracie's head and assure himself that she was comfortable. Turning back to McDougal, he said, "Pardon me for going on. It is just that I'm enthusiastic about our work. I trust you have not found me a bore."

"Indeed not," said the pastor. "Your ideas are a revelation to me. I confess that you make a case I can understand with respect to animal and plant breeding. It's just that when discussing human traits, I wonder if there is not a good deal more to it."

"Like what?" Nigel asked.

"Well, for example, in my work I find that problems in the family can often be attributed to life circumstances, misfortune, as it were. Let me give you an example. Recently one of my congregants came to me in crisis. His wife of thirty years died after being in poor health for nearly a decade. Through no fault of his own he became unemployed. He was a harness maker. Although I have known his family — a good family — for as long as I have been in service to the Lord, his abrupt turn to alcohol could in no way be attributed solely to inheritance of family traits. In fact, I know of no alcoholism in that family. It seems to me that his circumstances, rather than heredity, had more to do with it. Poor fellow. Though not to be excused, drinking was his way of coping. Allow me to give you another example. One of the poorest families in my congregation, the mother and father of whom are of limited intellectual capacity as you say, has a son of the most remarkable talents. The young lad excels in his school work and is an exemplar of all that we hold dear. How then can it be said that such intellectual capacity is inherited?"

"We do not doubt that there are irregularities such as you cite. We see them in our work occasionally. However, across large samples, we see a high degree of correlation between intelligence of the siblings, and between the siblings and their parents. In fact, we can measure the degree to which they align," Nigel replied. He once again became more animated in the discussion. "In fact," he went on, "our research shows that — how shall I say it — some families just have many more 'circumstances' as you call it." Sensing the conversation had taken a turn, Gracie rose up to look quizzically at her father, then settled back into her book.

"Forgive me for saying this," McDougal added, "but it seems that your emphasis on perfecting human traits places a higher value on only those that seem to you preferable. The suggestion seems to be, then, that some people are inherently better than others. I am reminded that our Bible says: 'For by Him were **ALL** things created that are in heaven, and that are in earth, visible and invisible....'"

Nigel interrupted and finished the verse. "... 'whether they be thrones, or dominions, or principalities, or powers: all things were created by Him, and for Him.'"

McDougal smiled. "I see you know the Bible. Anyway, as to my point, might not the soul of each man be pleasing to the Lord? Of course you know that God also said, 'Let Us make man in Our image, according to Our likeness....' And did Jesus not draw the poor, the impoverished and the afflicted to himself, while often rejecting those who had the most advantage? Yet, he did not heal everyone in this life."

"I think you would agree that we have the God-given capacity to think, to reason and to solve the problems which confront mankind. For our society to prosper, for the people to prosper, it is well that we should find ways to improve the lot of man. If that means we use selective breeding to eliminate disease, to end suffering, or degeneracy, or what have you, if we can make man better able to handle the complexities of modern life, wouldn't that be a good thing? Certainly you must agree at a minimum that intelligence confers advantage. Shouldn't we seek such advantage for everyone? Weren't we imbued with the capacity to reason and to manage the world around us? One cannot deny there are many problems in society. And one cannot deny that sometimes those problems continue through generations. It would be irresponsible for us not to take whatever steps we might to eliminate those problems."

"And just how do you propose to do that?" McDougal asked, still patiently engaged with what had caused him great inner turmoil.

"Now there are those who favor negative eugenics ... through various means ... that is, actively discouraging or preventing reproduction by the unfit. First, let me say that I am not a negative eugenicist. In any event, it starts with education," Wellbourne answered. "People must be given to first see the problem. Then, we need to explain the science of inheritance. Finally, we, and I mean the government, need to take steps to encourage reproduction by the fittest in our society, and discourage the same among those least likely to carry desirable germ plasm. This would be done over time, and with the greatest degree of sensitivity, of course. In the meantime, we must continue our scientific work to determine which are the traits most likely to be passed down through natural means. Oh, and as to the government, it is quite clear that the hard separation of the classes, particularly in my own country, serves as a barrier to inter … em … intermarriage between the most able. Using your example of the talented boy in a lower intelligence family — imagine if he was to marry someone of the finest breeding in America, something not now altogether likely. Wouldn't they produce better stock than if they were limited to marrying within their own class? I would suggest to you that the capitalist system is an impediment to human progress."

"There is neither Jew nor Greek, there is neither bond nor free, there is neither male nor female: for ye are all one in Christ Jesus," the pastor replied. "In God's eyes they all have worth and exist for His glory," the pastor noted.

"Exactly," replied Nigel.

McDougal was certain now that Wellbourne had missed his point.

"As I have discovered in my exploration of the ship, you can imagine that the best from among those in steerage might well be the intellectual and moral equal of any of those in first class. Wouldn't it be better, therefore, if we did not have such class distinctions? Wouldn't it be better if we had just one fare class aboard this and every other vessel?" asked Nigel.

"Umm," was his companion's only reply; he thought to himself that Nigel must have made the decision to purchase second-class accommodations rather than less expensive steerage tickets for some reason. Perhaps, McDougal thought, he did not want his child intermingling with their kind. Despite his expressed preference for a classless society, or at least one which actively promoted social mobility, perhaps Wellbourne in reality had embraced the status quo. Given his not quite superior station, he was at least comfortable. There were many more much less so.

"It is getting late," McDougal observed. "I must be getting back to freshen up for dinner. I have learned much from our discussion, and wish the best for you in your work," he added.

"Likewise," replied Wellbourne. "I have enjoyed this time immensely." He stood to shake the pastor's hand. "Come, Gracie, we must see to Mother. First, we need to stop off and order tea."

"It was a pleasure to meet you, Reverend McDougal," Gracie said enthusiastically as she waved goodbye. "Come, Father," she said, grabbing her father's hand. "We must remember to order a scone for mother, too. She is eating for two now."

When Nigel and Gracie arrived at their cabin, they found Gemma brushing her hair at the washstand.

"Feeling refreshed, dear?" Nigel asked.

"Yes, darling," she said. "I think I must have needed the time as I slept quite soundly. How was your walk?" she asked, turning to Gracie.

"It was grand," Gracie responded excitedly. "Papa and I saw gulls and many sea creatures, didn't we, Papa?"

"I am pretty sure we did, Gracie. I'm pretty sure we did," he replied.

"We also met a nice man in the lounge. He and Father talked a lot," Gracie added.

"That sounds nice. And what did you do while they were talking?" Gemma inquired.

"I read," Gracie answered. "Oh, and the man said I was very smart."

"You are, dear. You are smart, but one must not talk of such things. It is not nice to talk about yourself in such a way. It is best not to speak of your own attributes. It is much better to be modest," Gemma said, reproving Gracie. "Anyway, that is not why we love you."

"But Father does — talk about intelligence, I mean. That's what he and the man were talking about, wasn't it, Papa?" Gracie asked, turning to her father for support.

"Well, darling," Gemma interjected, "it is his work. Still, you make a good point. He risks becoming a frightful bore if all he talks about is his work. Isn't that right, my love?" she asked, turning to face Nigel.

"I suppose you are correct once again, dear," he said. Hearing a knock at the door, and grateful for the intrusion, he turned toward the door. Over his shoulder he added, "I shan't bother you any more with such talk, at least for now. Shall we have some tea?" he said as he opened the door to the steward.

After they finished their tea, Nigel suggested that it was perhaps a good idea for Gemma and Gracie to do a bit of exploring. He reassured them that he would not get into any mischief. It was just that he had to do some work-related reading. Just before leaving London, he had been in receipt of the Report of Proceedings from the 1st International Congress on Problems in Eugenics, which included the text of a speech about some of the most recent research coming out of the Cold Spring Harbor lab. Naturally, Nigel wanted to be fully informed about their most recent work before he was to meet with Davenport, the head of the lab.

Both Gracie and Gemma gave Nigel a gentle kiss on the forehead before hastening out the door to enjoy the late afternoon light and fresh sea breeze. Meanwhile, the dutiful if undemonstrative Wellbourne settled in for an afternoon of heavy thought. After all, Nigel Wellbourne was a thinking man, and in that, he thought himself unlike most men.

(*Something appears to be bothering you. You seem a bit off. You don't sound yourself. Perhaps you should rest and let me tell the next part.*)

[**I am feeling a might barmy, out of me head of late. This first part always brings up some ill feelings, and anyway, I've got quite a bit on my mind just now. If you insist, why don't you go ahead and continue for a while?**]

CHAPTER V

WHILE A YOUNGER NIGEL was first donning a clerical collar, Charles Davenport was in the process of soliciting funding from the Carnegie Institute of Washington. The grant was intended for a newly established eugenics laboratory located near Cold Spring Harbor, New York. Around the same time Davenport, who had for several years been receiving frequent, unsolicited correspondence from someone not then personally known to him, arranged to meet the eager writer. Bored with teaching science at the Kirksville Normal School, Harry Hamilton Laughlin of Missouri was the correspondent. After nearly three years of exchanging letters about research on chicken breeding and race-horse pedigrees, Davenport's interest was sufficiently piqued that he offered Laughlin a job. Davenport's decision was based primarily on the unrestrained enthusiasm Laughlin exhibited in trying to discover the biological rules by which desirable characteristics might be selectively bred into any given species. Of course, Davenport's primary interest was in humans, not domesticated animals. He could see that Laughlin had the necessary interest in inherited traits, an interest that could easily be redirected to humans.

When he was hired, Laughlin lacked the requisite education to attain credibility as a research scientist. His education had only prepared him to be a science teacher. His academic record was at best competent, certainly not remarkable. More relevant to Davenport was

Laughlin's keen interest in the subject of Mendelian inheritance, and his dogged determination to make a mark in the field. That he had experience as a school principal and later a superintendent persuaded Davenport that Laughlin could be expected to have the administrative and supervisory skills to oversee the work of others.

In spite of the limitations of his early educational background, over the course of his employment at the eugenics lab, Laughlin was to achieve a master's degree in biology and later a doctorate in cytology from Princeton. These achievements, along with his many work accomplishments and accolades, were to validate Davenport's decision to hire this zealous scientist-in-the-making. Laughlin started work at the lab in September 1910. He was from the very beginning of its funding the superintendent of the officially named Eugenics Record Office (ERO), with day-to-day responsibility for the office staff as well as recruitment, training and supervision of field-workers.

When in the late summer of 1912 Wellbourne arrived at the ERO, he was not known to Harry Laughlin. At the time, Laughlin had already hired dozens of field-workers, mostly young women. Unfortunately, more than one-third failed at their duties, and either resigned or had to be let go. For many, the reason had been a reluctance to pry into unsavory family traits. For others, it was their inability to grasp the detailed coding requirements of family pedigrees. In either case, the work of the lab could not proceed at the pace that was planned until the problem had been corrected.

"Good afternoon," Nigel said as he approached the receptionist located just inside the front door of the mansion that housed the ERO. "I am here to see Charles Davenport," he added confidently.

"Is he expecting you?" she replied.

"I believe so," said Wellbourne with assurance.

"I have no record of an appointment for him at this hour," she said, shuffling through her papers. "And whom shall I say is here to meet with him?"

"My name is Nigel Wellbourne. And this is my daughter Gracie. As we are just getting settled at a boardinghouse in town, my wife was not able to care for our daughter at this time. I hope you'll excuse this irregularity." The receptionist nodded curtly to the child, obviously perturbed at her presence and Nigel's unexpected intrusion.

Noticing the receptionist's displeasure, Nigel said to Gracie, "Why don't you sit over there by the window? I shall not be long."

"Sir," the receptionist said, "exactly what business do you have with Dr. Davenport?"

"He and I became acquainted in London, where I was in the employ of the eugenics lab at University College. We have exchanged letters in the intervening months. Most recently I advised him of my pending arrival in the States. At our meeting in London, he suggested that I should consider coming to work for him, a fact confirmed in my most recent letter a fortnight ago. I believe Karl Pearson has also written to him about my coming."

"I do not recall any such correspondence from London," the receptionist said, now raising her voice in mild indignation.

Hearing the commotion in the entryway, Harry Laughlin appeared from what must have once been a parlor. After looking Wellbourne up and down for a moment, then glaring over at Gracie, he said, "Good day, sir. What seems to be the problem, Miss Chalmers?" Then turning to look disapprovingly at the child once more, he inquired, "Is the child yours?"

Before Nigel could answer, the receptionist interrupted, "This gentleman says he has an appointment with Dr. Davenport, but I have no record of it."

By then, Gracie had slipped away from the front window where she had been seated, and quietly walked into an adjacent room, where several young women were fully absorbed in their labors. While Nigel, Laughlin and Miss Chalmers were sorting out the misunderstanding, Gracie began to speak to one of the young women.

As they were talking, Davenport had quietly entered the workroom through a back door. Noticing the child, he walked up behind her to see what was going on. A child at the ERO was uncommon, certainly one without adult supervision. Neither Gracie nor Davenport's employee saw his approach.

Gracie asked the young worker, "What are you working on?"

"Oh, it's just some calculations," she replied.

"What about?" Gracie asked.

"Dear, it is hard to explain, but it's mostly about the percentage of — of a condition of a certain type in a group of people."

"A condition?" Gracie asked.

"Yes, I'm calculating the prevalence ... er, the percentage of a ... ah, a kind of limitation in certain people," the young mathematician responded.

"Percentage! I know how to do that," Gracie exclaimed. "Let me see," she said. She looked over the young women's shoulder. "Oh," she said, "I think you might have made a mistake. I think you divided wrong."

The young woman checked her work, and indeed found the error. "Why yes, dear, you were right. I did make a mistake. Thank you.

Say, how old are you?" she asked as she turned to look into Gracie's face. With that, she noticed Dr. Davenport, who was standing behind them, presumably having witnessed the whole conversation.

Sensing the young worker's embarrassment, Dr. Davenport laughed and said, "Looks like we might have found ourselves another mathematician. Never mind, Miss Davies, mistakes happen. I'm sure you would have found it. Say, whose child are you, anyway?" he asked, stooping down to Gracie.

"That's my father over there," she said, pointing to Nigel, who was doing his best to explain his situation to the receptionist and Harry Laughlin.

Dr. Davenport straightened and walked over to the gaggle. He immediately recognized Wellbourne, but could not recall his name. "Good day, sir," he said, extending his hand to Nigel. "You are familiar, but forgive me; I don't recall the name."

"Nigel Wellbourne. Of the London ERO. I worked for Karl Pearson. We met ...."

"Of course. Of course," Davenport replied now making the connection. "You recently wrote and said you might be coming."

"Indeed. I wrote again just two weeks ago with my expected arrival date. I suggested that we might meet today. In fact that is why I am here now," Nigel said.

"Perhaps you just missed my telegram in reply. In any event, I am glad you are here. Welcome, and do accept my apologies," Davenport offered. "Would you like to come up to my office and we can chat more?" Knowing that Laughlin would be uncomfortable with the child left downstairs, he said, "Oh, and do please bring your daughter along. I would like to speak with her as well."

"Thank you," Nigel said graciously, " I shall. Thank you." He reached for Gracie's hand and together they followed Davenport up the stairs to his office, which once had been the home's main bedroom.

"Please make yourselves comfortable, and do forgive us for the somewhat inhospitable greeting in our front office. Had we known your exact date of arrival, I'm certain that you would have had a much more welcome first encounter. Between you and me, our receptionist Miss Chalmers is, given the nature of our work here, a valued but sometimes overly officious gatekeeper. As for Laughlin, being childless himself, he lacks the sensitivity that a father naturally develops. Please do accept my apologies."

"I understand completely, and accept mine as well. It would have been better for me to have sent a telegram on my arrival in New York," Nigel replied in kind.

"Oh, say, Gracie, if you like, you can find some of my daughter's books on the shelf next to you. Please help yourself. Your father and I will just be having a brief conversation. When we're through, perhaps you would accompany me to the kitchen for a snack, and I can show you around our grounds. We've got an old groundhog living out in the garden. He's a fat but amusing sort, and quite good fun. If you're game, he might even be willing to eat out of your hand."

Gracie looked over her shoulder while perusing the books, and smiled broadly at him.

For the better part of an hour, Nigel and Charles exchanged thoughts on the progress of their respective eugenics labs, all of which gave both of them confidence that Nigel could fit nicely at Cold Spring Harbor. Their respective work was in many respects quite similar, though through Davenport's initiative, the work in America had grown vastly more expansive. Having secured ample funding from a number of wealthy benefactors and the Carnegie Institute in Washington, a great many field-workers had been quickly hired, such that the task of training and supervision had overburdened Laughlin. Given Laughlin's infectious and often loquacious enthusiasm for the aims of the eugenics movement, Davenport concluded that Harry might be better utilized in the essential work of promotion, especially among the thought leaders of the day. Early on, Davenport recognized that progress could not be made unless society at large became willingly participants. Eugenics societies were proliferating and would need support. Outside experts could stimulate interest by giving talks on the most dramatic results of recent research activities. It was imperative, too, that doctors of medicine and social reformers, even ministers of God be recruited to the movement, as they were often among the first to encounter the persons whose limitations gave rise to its inception. Newspapers would be useful, so articles and press releases would have to be written. Laughlin had shown promise in all of these areas.

[To achieve desired ends, zealotry depends on many offspring.]

(Yes, and zealots often come to undesirable ends.)

"Wellbourne, it occurs to me," said Davenport, "that you have just the kind of skills we're looking for to take a leadership role in guiding the work of our field force. As I mentioned, we have been burdened by unexpectedly high turnover among this group. Perhaps it is a function of hasty selection. It's also a matter of giving them the proper training and supervision."

"Yes, of course. That sounds like something I should be able to get my arms around," said Wellbourne.

"No doubt," replied Davenport. "I'd like to begin concentrating Harry's time on supporting the organizations that call on us for direction and guidance. Universities, physicians, eugenics societies, and the like. As you know, our work is not merely research, but it includes by necessity the transformation of societal practice when it comes to marriage, physical separation of the impaired, and that sort of thing. In no way can we count on government mandates to make the changes we foresee. Any success we might have depends on voluntary compliance, and that will take considerable social change to have any effect. Of course, if we could in someway compel compliance with standards of eugenic marriage and various forms of segregation, we might see an improvement in the human stock in as few as fifty years. Anyway, that is my pipe dream."

"Yes, of course," Nigel said. However, he was not yet prepared to accept the idea of requiring eugenical marriage, or physically separating the physically and mentally impaired. Under the circumstances, he knew that he'd best keep his reservations to himself.

Davenport stood, then said, "I think that before we proceed, we need to make sure Harry is on board. It's a team effort, you know. If you are in agreement, I'd like you to spend a bit of time with him this afternoon. I want him to have as good an understanding of your background as I do. Then, if all of us are agreeable, we can have you start just as soon as you are settled."

"That sounds fine," Nigel replied. Though he had no reason for certainty about the outcome of this first meeting, he was becoming increasingly confident that a suitable place might be found for him in the organization. It's just that he did not expect it to happen so fast, and in no way did he assume that he would be given such great responsibility so quickly. It wasn't surprising, then, that Wellbourne was buoyant as he descended the stairs to meet with Mr. Laughlin.

"Harry," Davenport began, "Nigel and I have had quite a good talk. I can see how he might easily fit with us. In fact, he brings with him almost exactly the kind of experience doing family research we have been doing these last two years. He's even been published. With your concurrence, of course, I'd like you to spend some time with him to see if you agree. Then you and I can talk about whether and how we might divide some of our duties. I can share my thoughts on several new projects I'd like you to take on."

"Certainly, Charles. That sounds good. Come, Mr. Wellbourne, why don't we step back into my office? Oh, will the child be joining us?" he finished with a thinly veiled attempt to hide his contempt for the presence of a child.

"No, Harry," Davenport laughed. "Gracie and I will be grabbing a snack, then seeing to our friend, Mr. Groundhog."

While Davenport and Gracie enjoyed their time together, Laughlin and Wellbourne used their time in more serious pursuit. Nigel seized the lead and began to speak enthusiastically of his relevant experience. He was given only enough time to provide a general outline of his work in England before Laughlin interrupted. Laughlin was the type of man who had much to say, so it was inevitable that he would do most of the talking. As he spoke at length of the lab's initiatives and his own ideas for the future direction of the movement, he could see that Nigel was an attentive listener, and seemed to accept those ideas. In the end, that is what mattered most to Laughlin.

It became clear in the first half hour that both had much in common, at least philosophically, if not with respect to their regard for children. Without dwelling on it, but owing to his occasional passing remarks, Laughlin let it be known that he and his wife Pansy neither had nor desired children. As he said, they preferred instead a life in devotion to each other, and of course his work.

[A clear case of reaction formation, as we have seen among many who have committed grand misdeeds throughout history. I'm certain Freud would agree.]

(My turn. Now shush!)

Aside from these personal revelations, Laughlin shared a bit more of Davenport's personal philosophy on eugenics, a philosophy he asserted all employees were expected to embrace. Unlike Pearson and Galton, who by then accepted that the mechanisms of inherited characters were vastly more complex than previously assumed, Davenport had asserted that it was instead simple Mendelian "unit characters" that determined inherited traits. Of course, Laughlin emphatically agreed. His position was given with such forcefulness that Nigel was left only to nod and remark that he understood. Based on their work in England, which by then marked a well-known disagreement with the American's view on the subject, Wellbourne had concluded that more evidence would be required to accept such a simpleminded proposition. Certainly, he had thought, eugenic science would need to progress much more fully for such a rigid and simplistic position to be accepted as the final truth. He could not help but think that it was so typically American to reduce the complexity of nature to some childish level of comprehension. Perhaps, he sarcastically thought, it was a function of their less developed minds, and that in itself was proof of a weakened, capitalistic American germ plasm.

Beyond explaining Davenport's views, Laughlin used the occasion of their first meeting to emphasize his own personal philosophies of eugenics. Conceding at first the beneficial effects of some immigration, such as Nigel himself represented, Laughlin made it abundantly clear that he had found considerable problem with unrestrained immigration, particularly from eastern and southern Europe, although he had little data to support that view. He said that America, like England, was being brought to an incipient protoplasmic decline. These last few years, the surge in immigration, especially from those regions, had caused considerable social strain and attendant suffering. Such conditions, he asserted, were clearly observable in the failure of recent immigrants to adapt to American ways.

Laughlin dominated the discussion, but by the end of the interview the two did find time to exchange views on the nature of the work and fixes to the problems with the field staff. Together, they concluded that Nigel could most certainly be of service in support of this expanding initiative. The hour had gone by remarkably fast for Laughlin, if not for Nigel, who found Laughlin pedantic and sometimes dogmatic. Fortunately, they were interrupted by the return of Davenport and Gracie.

"So, Harry," Charles began, "how did it go? Did you two reach some accord on how we might think about bringing Nigel into our organization?"

"Yes sir, we had quite a good talk. I've learned much of Wellbourne's background and agree with you that he could be an asset — particularly in helping with our field staff," Harry replied.

(*You seem about to say something.*)

[**I can state unequivocally that Mr. Harry Laughlin was not without guile in the way he had seized an opportunity to free himself from the burden of day-to-day accountability. I know for certain that he had always retained grander ambition.**]

(*You must be feeling better … and of course no one would dare to contradict you!*)

[Is it sarcasm for you, then? Perhaps it is best that I get on with telling the story now.]

"Good, good," Charles enthused. "I'm wondering when you might be able to start, Nigel."

"I could start immediately but for the fact that we have not as yet found suitable housing. For now, we are temporarily in a boarding house. It's over on Sheep Pasture Lane, I think, just up the same road we came down to get here — 108, I believe," Nigel nervously said, offering unimportant details.

"Right," replied Davenport. "I know exactly where it is. Harry, I'm wondering if you would be able to take Nigel and Gracie to their boarding house. Then, if you are able this afternoon, take them into Cold Spring Harbor to locate something more permanent."

"Yes, I would be pleased to do that," Harry said. "We can use the time to talk more about priorities and how we might get him started in the work. That will give me the opportunity to meet your wife as well -- that is, if she's available."

"We have a plan," said Charles. Turning to Gracie, he said, "Well, young lady, it has been my pleasure. I do hope, once settled in your new home, that you will come to see us again very soon. Perhaps we can train Mr. Groundhog to do some tricks. He's a very smart fellow, you know."

Gracie smiled and shook Charles' hand, then reached for her father's hand. They all shared a laugh and went outside to say their goodbyes. When Harry went around back of the mansion to get the automobile, Nigel approached Charles, and spoke in a low voice. "Mr. Davenport, as the trip over has put a temporary strain on my …."

Before he could finish, Davenport removed his wallet from his coat pocket and offered all of the cash it held, some $50. "Say nothing of it. Consider it a goodwill payment. We look forward to your joining us, and know you will make a fine addition."

"Thank you, sir," Nigel replied, trying to hold onto a vestige of dignity.

Observing the transaction, Gracie knew that her father had been compelled to submissiveness through his own intemperate actions. She understood that the journey to America had strained their financial resources. A child, especially one as clever as young Gracie, could sense tensions over money. Davenport's gesture, or more precisely her father's predicament, made her uncomfortable. But she did not let it show in the same way as her father, who shuffled his feet in the gravel while looking at the ground.

While Nigel and Gracie were away, Gemma had used her time to freshen up their room at the boarding house. After a lunch of soup and sandwiches with a few other boarders, she retired to her room for a brief rest. Having accepted an invitation from Ardis Gilroy, the homeowner, to join her for late afternoon tea, she went back downstairs to the dining room. Nigel was not expected until right before supper, so Gemma would use the time to find out more about the area and the Eugenics Record Office from Mrs. Gilroy.

"How many years have you lived in the States, Mrs. Gilroy?" Gemma inquired while sampling the warm cakes that Ardis had just pulled from the oven.

"Cork and I — that's what I call Mr. Gilroy — we have been here going on 20 years. We find it to our liking. It reminds us of home — at least the sea air does," she answered in her still heavy Irish accent. "It's quiet, and safe. We've got a good parish priest, bit of a rogue at times, but he's got a clean heart. Anyway, this is our home," she said.

"Yes, I think it's lovely. And the air is so much cleaner than in London," Gemma added.

Ardis reached over and placed her hand on Gemma's forearm, then said, "I want you to know that your family is welcome here. And if you be needing anything, anything at all, all you need do is ask. We've not had so many young people around. Cork and I welcome the company, especially such fine people as yourselves."

"Thank you, Mrs. Gilroy," Gemma blushed.

"Oh," Ardis said, "you must call me Ardis. It's a small town, and we'll all be getting to know each other well enough," she added.

"That raises a question, if you don't mind my asking," Gemma said. "What do you know about the people and the work of the research lab? All Nigel has told me is that they look into family histories. He has told me only a little bit about Mr. Davenport. I have heard no one else's name mentioned."

"I can't tell you much more than what you already know," Mrs. Gilroy replied. "Harry Laughlin, he's the superintendant there. Harry Laughlin and his wife Pansy are all right, I suppose. They moved here from the Midwest not that long ago. No children … just the two of them. Mr. Davenport — he got the place started. He's nice, people say. He's up there in society. Meets with a lot of money people. He's gone a lot, too. I guess Mr. Laughlin runs the place day to day.

"What's he like — Laughlin, I mean?" Gemma asked, probing deeper, knowing that Nigel's employment stability would depend on the quality of the relationship with his boss.

"Can't say much, as I don't know him that well. People say he's quite religious. Maybe thinks he's a little better than some. Some folks say he can't have children. That's why he can be somewhat of a stick in the mud. I'm not sure. I've only spoken to Pansy once … at the market.

She seems nice. But she seems to kowtow to her husband. Once I saw him come in the store while she was shopping, and when he beckoned, she dropped what she was buying and headed out the door in a big hurry. That's about all I know."

"Anything else?" Gemma asked.

"Well, like I said, they don't have children. Once I saw him come out of his home and shoo the local children ... wee ones ... off his lawn. Scolding, you know. Seemed a bit harsh if you ask me. Anyway, people here mostly tend to their own business. I suppose I should as well, and not gossip so much. Father says it's a sin, but I've got to admit, stories are just sometimes too ripe not to pass on," she laughed.

Gemma nodded and laughed along with her. "I know what you mean, Ardis. And I have to agree. Anyway, if Nigel is successful in getting the job, I guess we will get to know Mr. Laughlin soon enough."

"Tell me, deary, how long will you be staying with us?" Ardis inquired. "No rush to leave, though. You're certainly welcome here as long as you like."

"Thank you, Ardis. I expect that Nigel will want to find a house to rent just as soon as we are able. He likes to have a study, his quiet place, where he can work in the evenings. If possible, we would like to have a separate room for Gracie and the baby. I school her now, so the extra space would be nice. Three or four rooms would be ideal."

"She seems such a bright child. I expect that she must take after her mother," Ardis gushed. "Anyway, I know someone just into town. They've a small cottage. Been talking about renting it out. Give me a day or so, and Cork and I can pay them a visit. Perhaps we can arrange something. I know you would find it to your liking, and it is still very close to the lab. Just about as far from it as we are here, and 'tis closer to town for you, so you wouldn't have much walk to the market."

Gemma smiled broadly and patted Ardis on the hand. "You've been so kind, Ardis. I don't know how we can repay you. If it works out, we must have you and Cork over for dinner some time. Oh, I think we are going to like it here in Cold Spring Harbor. I just hope things work out with Nigel and Mr. Davenport. Oh, and Mr. Laughlin, too."

CHAPTER VI

As Harry Laughlin prepared to depart the ERO office with Nigel and Gracie, Gemma and Ardis were joined at the boarding house dining table by Mrs. Maria Gulotti, another boarder. Maria and her husband Giuseppe had immigrated to the United States five years earlier. Owing to his experience in the family business in Sicily, Giuseppe opened the town's second and more successful fresh fish market. Maria gave birth to their first child not long after the market opened. Ironically, having fled Sicily to escape mafia violence, Giuseppe was killed in an armed robbery. Lacking resources to continue the business, and with a small child, Maria had been taken in by Ardis and her husband. When her child was old enough to take solid food, Maria began to work in Ardis' kitchen in exchange for her room and board.

Nigel, Gracie and Mr. Laughlin arrived at the boarding house just as Cork was arriving from town. The men found the three ladies at the dining table laughing conspiratorially as if at some private joke. The ladies pretended not to notice the men now standing silently at the head of the table, as if waiting for some sign of deference. Cork cleared his throat to get his wife's attention. The women exchanged knowing looks with each other, and then stood as they regained their composure. When Maria saw Harry Laughlin standing to the rear of the other men, she slipped away into the kitchen, hoping her hasty exit would not be noticed. Gemma observed Maria curl her lips in a soundless snarl as she left. Ardis and Gemma remained standing as Nigel made the introductions all around before Cork invited them all to sit for a bit.

"Mrs. Wellbourne, it is a pleasure to meet you," Harry said. Gemma was certain that she caught his eyes looking toward her belly as he spoke. "Mr. Wellbourne has told me a bit about you, but I am anxious to learn more. Perhaps when you are settled, my wife Pansy and I can have you both over to get acquainted."

"That would be lovely," Gemma said, realizing then that Nigel's interview must have gone well.

"Gemma, dear," Nigel said, "Mr. Laughlin has offered to show us possible rentals in Cold Spring Harbor this afternoon, that is, if you are up to it."

"That's very kind of you, Mr. Laughlin," she said. "Though perhaps it's a bit late in the day, and it may not even be necessary."

"Why might that be?" asked Nigel.

"Ardis has a friend looking to rent a small cottage in town. It sounds as though it might meet our needs, and it's possible that it is available immediately. Ardis said that she will inquire."

"Well, you all have been busy here, I can see," said Nigel with a grand smile on his face. "And I've got some good news to share with you in turn," he added. "I've been offered a position with the laboratory, and plan to start Monday." There was such a spontaneous outburst of approval that Mrs. Gullotti peeked out of the kitchen.

Cork, who had been standing with his hand resting gently on Gracie's head, offered to go into town before supper and confirm the availability of the cottage, then make necessary arrangements. His active involvement moved the plan beyond what might have seemed only a polite gesture, and his offer to help added to the good feeling.

At that, Ardis stood and said, "You'll never plow a field by turning it over in your mind." At first no one was sure if her comment was meant to nag Cork, but then she added, "Excuse me, but I've got to be getting into the kitchen to check on the evening meal. Will you be joining us then, Mr. Laughlin?"

"No, thank you very much. I am sure that Pansy has dinner waiting. Now, if you all will excuse me. It has been a pleasure meeting you, and I look forward to seeing you on Monday, Nigel. Perhaps we can meet at 9:00 in my office and lay out a work plan for your first month. In the mean time, I hope your search for more permanent accommodations goes well. If I can be of any assistance, please let me know. Goodbye, then."

Cork showed Mr. Laughlin to the door, then returned to the table and encouraged Nigel and Gemma to sit. Ardis, who had gone into the kitchen to check on Maria, came out with a fresh pot of tea.

"This is nice," said Ardis. "I'm glad we can have a moment of quiet before the hungry hoard assembles. Now, Gemma, is it a doctor you'll be needing then?" she said in a motherly way.

"Do you know one?" Gemma asked.

"Dr. Abraham. He was the doctor for my last," she answered. "That was a while ago, but he is still around, and has quite the following even now, especially with all the newcomers. Best one for babies, they say."

From the kitchen came a disembodied voice, " *Si. È un buon uomo.* Ee's a good man."

"Yes," said Ardis, "a gentleman. Got old-fashion ways. But he's kind and will take good care of you. I can take you to meet him if you like," she finished.

"That will be very helpful, Ardis," Gemma said. "You've been so kind."

"Think nothing of it," Ardis replied. "Now, Cork," she said as her voice got more directive, "Shouldn't you be going now if you intend to be home in time for supper?"

"Yes, Ardis," he answered in mock submissiveness. "I'd best be on my way." He left the table, grabbed his hat at the door, and left for town.

"Mr. Wellbourne," Ardis said, "'tis wonderful news about your new job. And with a wee one on the way. So many good things happening for you. Cork and I are pleased for you all. As I have told Mrs. Wellbourne, you are of course welcome to stay here just as long as you like."

"Thank you, Ardis. You and Cork have been most welcoming. It has made our transition so much easier. We are very grateful for all you have done. I am sure we would love to talk some more, but if you will excuse us, please, Gemma and I have some planning to do. Gemma, we should maybe get out of Ardis' way. We can go freshen up for supper, and maybe talk about a possible move. Come now, Gracie. Gracie? Where is that child?" Looking around, he spotted her in the parlor, where she was curled up with a book. "Come, Gracie, we should leave Ardis alone so that she can get supper ready."

"That's all right, Mr. Wellbourne. She can stay right there if it's okay with you and Gemma. I'll keep an eye on her. Or she can help me in the kitchen if she'd like."

Gracie looked up from her book and smiled at Ardis. "I would like to help in the kitchen."

"Surely, child." Ardis said. "Come, then," she said as she beckoned Gracie.

As Gracie entered the kitchen, Gemma suggested to Nigel that he go on ahead upstairs and that she would be up just as soon as she finished helping Ardis clear the dining room table.

After Nigel left, Gemma began in a quiet voice, "Ardis, I couldn't help noticing how abruptly Maria left when Mr. Laughlin arrived."

"I didn't want to say anything earlier," Ardis responded. "Maria has had encounters with Mr. Laughlin before — at the fish market. She once overheard him remark to his wife that it was good to see Italians running a business, since so many of them were either incapable or were content to live on charity. I guess he thought the Gullottis wouldn't hear his comment, or wouldn't understand."

"He didn't!" Gemma gasped.

"Cork has heard it, too. Even going on about the Jews. Anyway, I mustn't gossip. Got to get supper on." With that, she lovingly summoned Gracie, and headed into the kitchen, Gemma shuffled along behind with hands full of cups and saucers. When they finished stacking the dishes, and while Ardis was busy helping Maria, Gracie turned to her mother and, speaking softly said, "I don't think Mr. Laughlin likes children. He was not happy to see me at the office, like Mr. Davenport was. Mr. Davenport was nice, but Mr. Laughlin was annoyed that I was there. He snapped at me for sticking my arm out of the automobile window. He said I might get it cut off if another vehicle came by."

"Well, maybe he was just having a bad day. Your father wasn't expected at their offices today, so his arrival probably got in the way of Mr. Laughlin's work. Anyway, I think we need to give him the benefit of the doubt, and you shouldn't repeat any of this to your father. He has enough on his mind," Gemma said.

"Yes, Mother," she said, then paused. "I like it here."

"I like it here, too, dear. Come upstairs when you are done helping Mrs. Gilroy, and please do everything she asks, won't you?"

"I will, Mother," assured Gracie.

Troubled by what she was hearing of Mr. Laughlin, Gemma headed up to the second floor. She knew that she must keep any reservations about Laughlin to herself, because she didn't want to arouse any doubt in Nigel's mind about his new employer. Her conversation with Nigel must instead be filled with their shared joy over his new job and a possible home of their own.

Cork came back just as the boarders were assembling in the dining room, and the Wellbournes were making their way down the stairs. He caught their eyes, smiled broadly and gave them a positive nod. At the bottom of the stairs he told them he had been successful in arranging for their rental of the cottage in town. They could take possession that coming weekend, which would give the owners enough time to freshen things up a bit. The cottage came furnished, so Gemma and Nigel would be spared the expense of having to acquire whatever furnishings they might scrounge until their financial condition improved. To their added delight, he said that rent would not be due until the end of the month, a concession which Cork arranged on the Wellbournes' behalf.

Positively stunned by more good news, Nigel and Gemma could only look at each other in amazement. Gemma gave Cork a big hug, and both she and Nigel thanked him profusely.

[**I know what you are thinking. Nothing provides a test like a run of good fortune.**]

(*Well, ...*)

[**You can be so cynical sometimes. Sometimes blessings are just that.**]

79

(But you must admit, blessings can also be a test.)

The next day Cork took Nigel, Gemma and Gracie to meet the owners. It was, as they all so enthusiastically affirmed, a perfectly charming little cottage. Gracie and Gemma laughed when they simultaneously pronounced it good, just as they had done in their cabin on board the ship to America. Owing to the dearth of their possessions, the move could be accomplished in just one trip. Later that Saturday, Ardis and Cork stopped by with a housewarming gift of fresh baked bread and a pie. Ardis said that she would come by midmorning on Monday to take Gemma to meet her doctor, and to show her around town. Subject to Gemma's approval, Cork would take Gracie back to the boarding house with him so that she could spend the day playing with Maria Gullotti's son Anthony, who was nearly the same age. That would give Ardis and Gemma plenty of time to explore the little town and do a bit of marketing. Situated as it was close to town, Nigel would use one of Cork's old bicycles to ride the few miles around the southern tip of the Inner Harbor to get to the ERO lab. While the weather was still good, he would not have to rely on other transportation. So it was, in the early fall of 1912, Nigel could see his life's quest for creating a more virtuous humankind was at last beginning to take form.

Monday morning, Nigel arrived at the grounds of the ERO office around 8:30. He lingered just outside the wide front gate to contemplate his good fortune. In his mind's eye, he could visualize that this Eugenics Record Office, this nexus of human reform, was to become, inevitably, a crucible of one of mankind's greatest achievements. Immobilized in thought as he was, he realized a passerby might have thought his behavior queer. At length he wheeled his bicycle up the front walk. Seeing no evidence of others present, he rapped loudly on the door. Miss Chalmers, who had arrived an hour earlier, opened it, offered a vague smile, and allowed him entry.

"As you can see," she said, "I am the only one here at this hour. Mr. Laughlin customarily arrives just before 9:00, as I'm certain he must have told you. You may wait in his office around the corner if you wish. If you'll be needing anything while you wait, please let me know." She sat back at her desk, glared at him briefly from under furrowed brows, and returned to the stack of paperwork piled before her. He did not register the meaning of her scowl as he paused to look back at her from the entrance to Laughlin's office. He was already thinking about how he might begin to impress Laughlin.

As was indeed his custom, Laughlin arrived five minutes before 9:00 A.M. Punctuality was to him among the most favored of man's qualities, and he insisted on no less in others. It was, in fact, the rigidity of this tendency that produced his abrupt command to Pansy in the market, the incident Ardis had once observed. Fortunately for Nigel, it was a characteristic he shared, at least the punctuality part, and it was one that had already been noted by Laughlin. Even before Nigel had produced any work, Laughlin deduced that Wellbourne was a man sufficiently fit for any duty he might be asked to undertake.

Harry seated himself behind his desk and gestured to Nigel to take a seat across from him. He carefully rearranged the papers on his desk so as not to be distracted, smiled broadly and once again formally welcomed Nigel to the Eugenics Record Office. After inquiring of Nigel's accommodations, and having assured himself that Nigel's family was settled in, he began to lay out a work plan for the ensuing weeks. The plan, a copy of which he gave to Nigel, spelled out Nigel's schedule for the next month. He and Nigel would meet each of the staff members in a group that morning. Nigel would be introduced as the new field supervisor responsible for hiring, training and oversight of the data collection effort. When he had become sufficiently familiar with that operation, he would be given responsibility for the office staff charged with compiling and coding records from the data collection effort. In the afternoon, Nigel would sit with the office team members individually to learn how records were compiled and filed. This would give him an opportunity to get to know the capabilities of each staff member. Days two and three would be spent assisting in recording of field questionnaires. By the end of the week, Nigel would be expected to be able to do his own compiling of records under close supervision by a more experienced staff member. The following week, Nigel would accompany Harry on a recruiting trip to Vassar College to select two new field workers, and continue assisting with record keeping in the lab. At the end of the first month, he would sit in on a training session for new hires, which Harry would conduct. Inasmuch as the work was so similar to what Nigel had been doing at Pearson's lab, the transition was expected to be easy for him, although he would have to adjust to slightly different forms and systems of pedigree charting and coding. Still, he did not expect the process to be so different as to cause him any delays. Rather, it was the manner in which Laughlin spelled out in minutest detail the work to be performed that presented Wellbourne with the greatest consternation. He had, by the time of his resignation in England, achieved a much greater degree of independence in his work. Perhaps, he thought, this was just Laughlin's way of ensuring consistency, and that when he had proven himself, Nigel would be given much greater latitude to work independently. These were, he told himself, minor adjustments required of any new employee, and in no way represented a problem that he would need to share with Gemma.

During Nigel's first week, Gemma and Ardis accomplished all that Ardis had promised. They walked about town, did some marketing, and took time to acquire a few household items that would personalize Gemma and Nigel's new home. Most importantly, Gemma met Dr. Abraham, who would attend the birth of her second child.

As Ardis had assured her, Dr. Abraham was both kind and gentle, and much more avuncular in his approach to patients than what Gemma had experienced in England. She and the doctor discussed her first delivery, Gracie's development, and both her and Gemma's general health. Somehow, the doctor developed a good understanding of his new patient, but the visit had all seemed so conversational. It was like Gemma was sitting with a favorite relative she had not seen in a long time. Their conversation touched on many topics unrelated to medicine, but that nonetheless conveyed much to the doctor about his new patient. It was a method he used with all of his patients to develop trust, and to glean information about their approach to healthy living. It was strange, though, that he wove into his remarks, at what were the most unexpected times, several of his own particular medical biases. During one pleasant strain of the conversation, he offered what was more non sequitur than a conversational divergence: "I do not believe in using forceps. They are, to my mind, cumbersome and sometimes harmful. Besides," he offered, "I have never quite gotten used to them." It was a statement that, for Gemma, was somewhat less than reassuring, and in no way flowed from his previous comment about the gardens of London. Sensing the need to fortify his patient for what lay ahead, he did offer that he liked to take things slowly, and not rush the process. Depending on circumstances, that also was for her somewhat alarming because she assumed that with birth pains it would be better to get things over with quickly, especially since Gracie's birth had been so difficult. Later in the conversation, while they were discussing Gemma's diet, he indicated that he preferred to not use chloroform unless absolutely necessary. This approach was necessary to ensure that the mother was active and alert in the birthing process. She agreed that was generally her preference as well. He concluded by saying, "Sharing my beliefs in these matters often induces patients to seek medical services from one whose views are more compatible with their own needs." His medical philosophies, if not idiosyncrasies, were offered even as he managed to carry on simultaneous conversations with Gemma and with Ardis. In reviewing Cork's health with Ardis, particularly his recurrent bouts with gout, he managed a fluid transition from speaking with Gemma to talking about a patient not then present. The conversational diversion allowed Gemma momentary relief as she was beginning to feel mild queasiness talking about forceps, chloroform and birth complications. Being able to divert her attention from the doctor permitted her to reflect on what a wonderfully perfect child Gracie had turned out to be. Of course all she wanted as a mother was a healthy baby, but she couldn't help but think how wonderful it would be to have another exceptional child.

The hour-long visit concluded with Dr. Abraham assuring Gemma that he was attentive to the need to practice listerian antisepsis, particularly given that hers would be a home delivery. This final conversational detour into his method of medical practice once more confirmed for Gemma that Dr. Abraham was, if nothing else, serious about his work. It was a view he was emphatic about. "Home births are preferable to the hospital, especially given the prevalence of infection in the latter setting." With a nod to Ardis, whom he asked to be available to assist at the appropriate time, the appointment was concluded. It was no small consolation to Gemma that at least she had come away with a plan for the birth of her second child.

On the way home, Gemma admitted to Ardis that Dr. Abraham's strongly held views were mostly comforting, and that she was grateful to Ardis for the recommendation. She could see that, though he was sometimes distracted, he was dedicated to his patients' wellbeing. She understood that she did not have the medical knowledge to judge his approach, though to her it seemed sound and practical. It was his confidence and reassuring manner that had most elicited trust, and she knew that in the matter of giving birth, this was the most important thing.

[**Doctors are exceedingly confident in what they know to be true. After all, they are men of science.**]

(*Sensing some irony there. You are feeling more like yourself, aren't you?*)

By the end of their first month in the United States, Gemma, Nigel and Gracie had settled into a comfortable rhythm. Nigel had already proven his capability to Harry, who could finally see a path to fully extricate himself from responsibility for data collection and record keeping, parts of the enterprise that were for Laughlin unfulfilling. Gemma had assured herself that she could manage their new home with the support of her new friends and Gracie's help, and that she was prepared to bring a new life into the world. And as for Gracie, she not only developed a joyful friendship with Anthony, but also found a new role for herself as her mother's great helper. Moreover, she would soon be the older sister to a new sibling. In less than a year, she would be ready to start school. Having been so well prepared by her mother, she would not in the least be afraid to take on these new challenges. For the Wellbournes, life was beginning to fall into place.

Ardis and Maria continued to be among Gemma's closest friends. Save for Lady Burgess, Gemma had never had such trusted friends in her life. They were at her side when she needed help around the house. They were invariably present for emotional support amd they were there just for tea and laughter. And because they were there, Gemma knew that she was home. When the time came for her body to signal in the most unmistakable way that their lives were once again to change in the most profound way, everyone she cared for would be there for her ... or almost everyone.

Nigel had, of course, worked up a detailed plan, or as he put it, a 'system' for the birth of his child. Owing to the fact that he would no doubt be at work, it would fall to Cork to fetch the doctor, because Cork had an automobile. Ardis would accompany Cork on the way to the doctor and be dropped off at the Wellbournes' to prepare for and assist with the delivery. Maria would manage the boarding house in her absence. After dropping off Ardis, Cork would seek out the doctor. Cork would then drop off the doctor at the Wellbournes', pick up Gracie, and return to the boarding house, where Gracie would reside for a few short days until Gemma had fully recovered from delivery. It had already been arranged that she would room with Maria and Anthony. Though disappointed not to be present with her mother for the delivery, everyone agreed that the delivery of a baby was not an occasion for the presence of a child. In the unlikely event that he was not at work, Nigel would was expected to absent himself from their small cottage so as not to be in the way. Every contingency was planned for. Everyone had an assignment and a place to be, so they all waited in eager expectation for the day when the 'system' could be initiated.

Laughlin had been called away to Wellesley, where the dean of students, a strong supporter of the eugenics movement, had insisted that a representative of the ERO teach a three-day junior/senior seminar on the principles of eugenics. Because the dean personally donated a portion of his savings to the ERO, and because the school had been a source of much needed field-workers, it was imperative that his wishes be accommodated. In Laughlin's absence, Nigel would need to be completely in charge of the lab. Laughlin might have chosen to relieve Wellbourne of this burden, given Gemma's apparent readiness to give birth, but Laughlin did not trust office work to proceed apace without close supervision. Alas, there was no other way, and everyone agreed that these circumstances were inauspicious.

Nigel's whereabouts, at least insofar as he was not at home, were of no concern to Dr. Abraham. In fact, it was to the doctor quite preferable that Nigel be elsewhere during delivery, and certainly not close enough to intercede should there be any difficulty. He found that fathers' fears during birth and delivery added to the tension the mothers already were feeling, only complicating difficulties that much more.

Then, Gemma's time came, or so it seemed.

CHAPTER VII

FORTUNATELY FOR ALL PARTIES CONCERNED, Gemma had merely experienced a false labor.

A day later, when with greater certainty Gemma's body signaled its actual readiness to give birth, her caregivers awaited the cue to assume their assigned roles. A pity it was that Nigel, who only knew that delivery was likely to be within days if not hours, had no knowledge of the immediacy of birth. He was, at the moment of Gemma's first crying out, sitting at a desk in the lab, there obligated to confirm mathematically that polydactylism ran in families. As carefully constituted plans have a way of going awry, he would not know of the delivery of his second child for many hours.

To a near silent symphony of pencils relentlessly scratching out labored calculations, the office ladies worked heads down, while Nigel was lost for a time in self-centered reflection. The regularity of office routine provided a lulling backdrop as he mentally retraced the arc of his life from schoolboy days through his time at University College, to Wycliffe Hall, then to his meeting Gemma at Letchworth. Ah, and the quince, oh, there was that luscious quince. Carried away by the memories, he thought on the chance meeting of Davenport at Pearson's laboratory. The sheer serendipity of it had in turn precipitated his journey to the United States. Despite the challenges he had faced along the way, he knew that he had been blessed in so many ways — with a loving wife, an intelligent child, meaningful work, and, of course, his own gifts. Yes, these were the blessings that accrue naturally, he thought, when one dedicates oneself to the noblest of pursuits, and takes advantage of all that one's innate talents afford. If only all men could share in such blessings, he ruminated. Lost in thought, he did not notice Miss Chalmers' looming presence.

(*Our poor Nigel ... lost in thought again. Perhaps I should say delusions.*)

[He might have considered them visions, though to me, they were not the equivalent of Daniel's.]

(*Maybe more like Nebuchadnezzar's.*)

It's funny about plans. They have a way of unraveling, even for those who are most rigorous in their preparation. So it was for Gemma and Nigel. While he had designed an elaborate system for the pending birth, and arranged proper roles for Ardis, Cork, Maria, Anthony, Gracie and the doctor, he had not thought through the contingencies. Cork was to have been the first domino to fall in a chain reaction that would sequentially move everyone into his assigned position. It's just that someone had to push the first domino. The boarding house was nearly a mile away from Gemma's residence, too far for Gracie to walk unaccompanied, so it was thought she could not possibly be the one to initiate the message that labor had begun. Nigel was at work, also a mile away from his home, and reaching the doctor, assuming he was even in his office, was similarly problematic. As Gemma and Gracie were home alone, it was conceivable for the delivery to occur before the doctor could be located. Such a happenstance was not completely irregular, though in a birth with complications the consequences could be dire. Besides, an unattended birth was unacceptable to Nigel.

(*To paraphrase something you have repeated often:* "*You can make many plans, but the LORD's purpose will prevail.*")

[Plans can indeed go awry. Nonetheless, "Fate" can somehow work things to advantage.]

Miss Chalmers interrupted Nigel's ruminations by passing along an urgent message from Mr. Laughlin, who requested that he be provided data contained in a recent report. Unfortunately, Nigel had taken the report home to read the night before and neglected to return it to the office. At Miss Chalmers' insistence, Nigel would have to go home immediately, collect the report and return to the office, where he would call Mr. Laughlin with the requested information. It was no small inconvenience, but owing to the fact that Nigel did not have a telephone at home, the demand would require him to immediately bicycle the mile home to retrieve the requested information.

Pedaling furiously, Nigel reached home, grabbed the report from his office, paused just long enough to look in on Gemma and Gracie, and then immediately raced to the door on his way back to the office. In the instant before front door slammed shut, Gemma shouted the phrase, "It has begun!" Had he lingered just thirty seconds, he might have been more certain of what it was that he heard. Her utterance was, of course, the triggering event that required someone to notify Cork, who in turn would use his automobile to locate Dr. Abraham. Pressed by the urgency of Laughlin's demand, in his haste Nigel had not at first registered what it was that he heard. By the time he correctly deduced its meaning, he was a quarter of the way back to the lab. Thereupon, he paused to reflect on the imperative of his boss's request weighed against the urgency of his wife's condition. It was for Nigel a matter requiring considered thought. Puzzling it out, he concluded at last that because no one else was available, he must first ride to the boarding house to notify Cork, and only then the mile to the office. By the time he got back to the office he was well lathered. It would take several calls to get through to Laughlin, who having received the requested information, added a series of additional reports from which Nigel was expected to extract summaries. When done, he had to call Laughlin once again to transmit the requested information. Adding to his misery, Nigel was under the impatient glare of Miss Chalmers, who by then had grown exasperated with his having tied up her desk and the lab's only phone other than the one in Davenport's locked office.

In his race to begin the search for Dr. Abraham, and with Ardis nowhere in sight, Cork neglected to bring Ardis to the Wellbournes' to assist with the delivery. At the time, she was out back of the shed preparing to cut the heads off two chickens, which were to be the evening meal. When the doctor was located and delivered to the Wellbournes', Cork dashed back to the boarding house to locate Ardis, who by then was up to her elbows in blood and chicken feathers. Meanwhile, Dr. Abraham found Gemma deep into labor, with contractions just a minute apart. In this mad rush, Cork, who had successfully located Ardis, forgot to bring Anthony along on the short trip back to the Wellbourne cottage. Though a minor part of the plan, it was assumed that Anthony's presence would provide comfort to Gracie, who of necessity would be removed from the Wellbourne residence. This was so that she would not be confronted with the discomfort of her mother during labor and delivery. No matter, as Gracie would prove obedient, albeit at first hesitant to leave Gemma's side. After dropping Ardis off, Cork took Gracie back to the boarding house, where per the original plan, she would stay for a few days, giving Gemma a chance to recover from the delivery.

It was early evening by the time Nigel finished his work for Laughlin. The office staff, including Miss Chalmers, had already gone home. Nigel bicycled home, frazzled by the day. There, he expected quiet repose with Gemma, and hopefully a new baby, free of interruption and intrusion. When he arrived, the house was quiet. Only the light of a small lamp shone through their bedroom window. Nigel supposed that Gemma would be resting quietly, and that Ardis would be in the front room resting. He parked the bike at the side of the house, and entered through the front door. Ardis was asleep in the front room. He quietly called out to Gemma. From the back room came a whisper, "We're awake."

Nigel, who had that day been totally absorbed by the necessity of Laughlin's unremitting requests, could only assume what had transpired at home. Gemma's labor and delivery proceeded without complication. The doctor was all that had been promised, and with Ardis's assistance managed the whole process with the greatest sensitivity. When the baby emerged, the child was cleaned and wrapped tightly in a soft blue blanket. Ardis had brought it along just for this purpose. It was the same one she used for the birth of her own children. She believed it would bring good luck for her friend. While the child was cleaned and checked over, Gemma waited expectantly for a chance to greet her new one.

Dr. Abraham was busy finishing the delivery process while Ardis cleaned and swaddled the child. With that going on, Gemma hesitated in making inquiry as to the condition or gender of the child. She was at the moment overcome with euphoria, and the feeling of release that accompanies childbirth. When the medical necessaries were concluded, Dr. Abraham handed the child to Gemma. "You have a son," he said. A tear ran down Ardis' cheek as she looked first to Gemma then to the doctor. Gemma tried to discern the message conveyed in their silent exchange. Dr. Abraham had acted differently now, dispassionately and coldly, not like the avuncular manner he had shown in his office during their first visit.

She peeled back the blanket to have a closer look at her son. At first glance, she told herself that he looked neither like herself nor Nigel, but she knew that was true of most newborns. Looking more closely, she thought she detected something different about his appearance: almond-shaped eyes, the shape of his ears maybe. But she wasn't sure. It is so hard to tell with newborns. As she held him, she noticed he had not moved much. In fact, he had never moved much in the womb. He didn't even cry out on delivery like most newborns, certainly not like Gracie. "Dr. Abraham," she asked, "is he alright?". He seems, well, different ... certainly different from Gracie when she was born."

"I'm afraid he is different," the doctor replied without emotion. "He's a Mongoloid." Before he could explain, Gemma interrupted.

"Are you sure?" she asked. I don't see how that could be." She was confused.

"I'm afraid it is true," he said.

She started to panic. "How can you be sure?" Gemma asked, convincing herself that Dr. Abraham was somehow mistaken. She remembered their first meeting, and how he insisted on clinging to some of the old fashioned ways. Maybe is was just wrong this time. "How can you tell? He looks fine to me." Her voice was raised in defiance. Normally self-composed, Gemma was starting to be afraid of what her baby's condition might mean. She'd had never had an encounter with a Mongoloid, let alone experience raising one with that condition.

"Gemma, I know this comes as a shock. Though these things are rare, they do happen, even to healthy people. See here, look, this crease that runs all the way across this one hand. This occurs in most Mongoloids. See how you have two lines. And the listlessness ... his floppiness, you might say ... you have no doubt observed this already.

You saw that he never made a sound when he came out. These things are how you can tell. In time, you will see that his facial features are not normal also."

As was his custom, his demeanor masked the gravity of his concerns when interacting with a seriously unwell patient. He did not want to cause alarm. Raising a Mongoloid child, he knew, would prove to be difficult if not completely unworkable, that is if they even survived early childhood. Most lacked hardiness, and were afflicted with other serious medical problems. In his view, it was best for families to immediately turn the care of such children over to institutions better able to care for them.

[New life is always a baptism — an emergence from darkness into light, from submersion to a gasping for air ... then a crying out in a piercing declaration.]

(Sometimes I think it must be a cry born of shock from the first encounter with both the sacred and the profane.)

Dr. Abraham lingered a while to assure himself that both his patients were stable, and to offer a few instructions. Then he walked back into town alone. Ardis remained with Gemma and held the child for a time before placing him in the crib alongside Gemma's bed. While Gemma and the baby slept, Ardis contemplated the troubles this child's life would bring, but assured herself that she alone, if necessary, would stand with the Wellbournes in the face of whatever might come their way. She would insist, if need be, that Cork would help, and together they would do whatever was necessary to make the Wellbournes' life as easy as they could. Before long, Gemma awoke, and Ardis gently handed the baby to Gemma for a feeding. Whatever limitations might have been attendant on a Mongoloid infant, taking nourishment was not one, at least in this child's case. Dr. Abraham had warned Gemma that Mongoloid infants typically had low muscle tone, so feeding might be a problem. He also warned that they were placid infants, and might need stimulation to stay awake during feeding. Consequently, he advised for now, more frequent feedings. It turned out that none of this was true for the Wellbourne lad. He was an eager and active feeder, though as far as frequent feeding, it would for him be a matter of necessity, as there seemed to be no limitation to his appetite.

It was during this first feeding that Nigel arrived home. When Gemma quietly answered his call, saying, "We're awake," Nigel paused in the bedroom doorway. Lit by a single bedside lamp, the scene was a tableau of peace and love. Gemma glowed beatifically with the new-born at her breast, swaddled in the light blue blanket that had been given as an offering. Ardis smiled and looked on from a chair in the front room as Nigel stood immobile. A broad smile eventually creased his face, and his chest swelled, when Gemma whispered, "It's a boy. You have a son, Nigel. We have a son." As sweetly as she said it, her countenance betrayed worry.

Nigel approached the bed as Gemma gently removed him from her breast. She wrapped him closely and passed him up to her husband, who bent down and cradled the boy gently in his arms and hand.

"You are a good father," Ardis said from the doorway to the bedroom. "You obviously know what you are doing. And you've got a fine wee lad, there," she added as Nigel pulled back the blanket to have a closer look at his son.

Nigel stared into his son's face, then looked back at Gemma. "Whom do you think he looks like, dear?" Even as he spoke, he sensed something was different. He looked back at the child, then Gemma. "He doesn't seem to favor either one of us." He could see fear in Gemma's face right before a tear rolled down her cheek. There was a moment of recognition, then confusion, as when you meet an old acquaintance whose name and history you can't quite place.

"Nigel, dear," Gemma whispered. "Our son has Mongolism. I'm so sorry."

Nigel stared at Gemma, incredulous.

"He is so sweet, though" Gemma said. I think we can learn to love him. We must try."

Nigel peeled the blanket back to look more closely at his son's face. Nigel was as stunned as Gemma had been earlier. He looked up at Gemma, then back to the child. He drew the child close, dropped his chin across his son's chest, and began to cry. And when his tears turned to sobs, he passed the child back to Gemma and knelt by her bed. He rested his head gently on her shoulder, and wept with her, but they were not weeping for the same thing. Hers were tears born of Nigel's doubt.

Ardis slipped away to the kitchen to make tea, and left the Wellbournes to comfort each other in private.

During the night, Nigel dozed intermittently at Gemma's bedside, while she and the baby slept. His fitful rest was too often interrupted by thoughts of the prospect of having to raise an imbecilic child. He could not conceive of how it was even possible for himself and Gemma to have such a child. Gracie was proof of the improbability of it. Weren't they themselves of the highest caliber, and hadn't they come from good stock? It was all so mathematically improbable, and so unjustified.

He awoke very early in the morning, and finding Gemma still asleep, crept into the second bedroom to lie down. Around 6:30 A.M. he was awakened by the sound of an automobile out front. A door slammed shut and there was a knock on the door.

A little earlier that morning, while still in bed, Cork heard the front door of the boarding house close. He rose quickly and went to the entryway to see who might have entered. Seeing no one, it occurred to him that he might just have been dreaming, or perhaps heard a tenant leaving to use the outside privy. Turning back from the front door, he saw Maria standing at the bottom of the stairs. In a somewhat frantic voice she asked if Cork had seen Gracie. She reported that she had heard her bedroom door close, and rose to check on Anthony and Gracie. Anthony was asleep, but Gracie was nowhere to be seen. Thinking that Gracie might have come downstairs, Maria decided to check further.

Without deliberating, Cork rushed out the front door to the porch, looked down the road toward town, then scrambled off the porch to check around back. Realizing that Gracie may have decided to walk home, he rushed back into the house, dressed, then raced around the house to the shed to get the auto and head toward the Wellbournes' cottage. A quarter mile down the road, and just out of sight of the boarding house, he found Gracie hastening on toward home. She got in, and the two drove on toward her parents, arriving around 6:30. Cork rapped on the front door, and was greeted by Nigel, to whom he delivered his daughter, saying that she missed her mother. Owing to the hour, Cork left and drove back to the boarding house.

"Did we have a baby yet, Papa?" inquired Gracie.

"Yes, dear, we did," answered Nigel, though his tone signaled concern to Gracie. "Why don't we go sit? I've something important to tell you."

"What is it, Papa? What's wrong? Is the baby all right? I want to see him."

"Mother and the baby are sleeping. We should talk first," he said, trying to reassure her. Sitting now with his arm around Gracie, Nigel said, "The child is fine. It's just that he has a condition"

"What condition?" asked Gracie, more curious than fearful.

"Well, Gracie. It's hard to explain, but he has something called Mongolism."

"What's his name?" asked Gracie.

Taken aback by Gracie's seeming inability to grasp the significance of what he had just said, Nigel recovered and pressed on with his explanation. "We haven't discussed a name. But, Mongolism is something very few people have. They do not get over it. What it means is that they learn slowly, and they look just a little bit different."

"Papa," started Gracie. "There is no hurry to learning," she offered as reassurance.

"It also means they cannot learn as much," he said.

"How much do we need to learn?" asked Gracie. The way she spoke was more a declaration than a question, and seemed to be jarringly presumptuous for such a small child. "I think we should call him Eugene," she said now, emphatically and with much enthusiasm.

Nigel was certain then that Gracie did not understand, but he did not want to disregard her childlike optimism. So he played along, "Why Eugene?"

"It sounds like a word that you use … eugenics. It makes me think of you, Daddy. Besides, I met a boy on the boat named Eugene, and I liked him. I like the name; don't you, father?"

"We'll see, Gracie. We'll see. We will have to discuss it with your mother. Anyway, I think they might be up now. I heard the baby cry."

With that, he lifted Gracie up off the settee, and the two of them walked quietly into the bedroom holding hands.

Gemma was nursing the baby when they entered. Gracie rushed to her mother's side and rested her head gently alongside her mother's. She leaned over and gave her baby brother a kiss, then looked up and smiled at her mother. Gemma smiled back and gave Gracie a kiss on the forehead.

"We should call him Eugene. I like the name very much. What do you think? Can we?"

Gemma patted Gracie on the head and said, "We'll see. We'll have to ask your father, but I do like the name." Gracie knew parents always say that, but in her heart she knew they would agree this time.

Later in the day, while Gemma and the baby rested, Nigel was making Gracie lunch when he tried once again to talk with her about the problem with the baby.

"If we keep him, the baby, I mean, things will be very difficult … for all of us," he said.

"They won't be difficult for me," Gracie said, sounding resolved.

"In time, you might see that adults will be uncomfortable around him, and children will make fun of him," said Nigel, now growing increasingly frustrated by Gracie's intransigence. "These children most often go to live in a place better suited to their care."

"Not in front of me they won't they won't make fun of him! Besides, he can't go away. He has to stay with us," she replied, now more obstinate than before.

Conflicted though Nigel was, he opted to drop the conversation, seeing that Gracie was not bothered by her infant brother's prospects. She had grown increasingly resistant to any intimation that there would be difficulty going forward. Nigel knew it would do no good to press the battle at this time. In the days that followed, he would make every effort to reestablish some semblance of routine that had just been interrupted by this new complication. At the time, he rationalized that he was only testing Gracie's resolve to live with a defective brother. In reality, he was testing his own resolve. Whatever joy he took from Gemma at her having borne another child, Nigel was quickly coming to the realization that in every way raising a defective child went against the core tenets of eugenics, and in no way could he expect to find acceptance among his colleagues at the lab.

Unannounced, Harry Laughlin called on the Wellbournes late the morning of the third day. Cork had told him of the child's arrival, so he came under the pretense of bringing a nourishing meal. Pansy, he said, had prepared it in the hope that it would make their lives a bit easier. He said that they wanted to help, and that this seemed to be the best solution. It was an awkward exchange, almost as if Laughlin had memorized his lines. Nigel did not offer to have him come in, which was clearly Harry's intent. Laughlin's curiosity was in no way masked by the obvious way in which he peeked around Nigel's broad shoulders, hoping to see into the inner sanctum. Nigel said that he'd love to have him in, but that Gemma and the baby were sleeping. Harry could no doubt hear Gemma cooing softly to the baby in the other room. It was the first of many lies Nigel would tell to conceal their family's circumstance. Nonetheless, hoping to buy time such that admittance might yet be gained, Laughlin asked, "What will you call him?"

"Eugene. We all like the name Eugene. Perhaps when Gemma is up to it, you and Pansy can come round and meet our new son." Nigel wanted to sound hospitable, but without making a commitment.

"It's a splendid name," said Harry. "It will suit him just fine. Perhaps in a day or so, Pansy and I can stop by to see how you all are doing, and meet him."

"Yes, that would be fine," he said. "If Gemma is up to it," he added. This time Nigel was sure that that Laughlin would interpret reluctance in the offer.

"Right," said Harry, "of course…if Gemma is up to it. Why don't you take the next few days and help out around here? We'll see you Monday when everything gets back to normal. I'm sure everything will be fine by then. We'll miss you in the meantime, but enjoy your time with family." He did not convey that he suspected anything was wrong, but Harry was a perceptive man. He must have somehow gotten wind of trouble. Nigel was sure of it. Either way, Laughlin could be counted on to sniff around to see what he could find out.

Shortly after Harry left, Dr. Abraham stopped by to check in on Gemma and the baby. Nigel and Gracie remained in the front room to give them their privacy. Dr. Abraham listened to their hearts, checked their pulses, and inspected Gemma's stitches. Putting away his instruments, he asked Gemma if he she had given any thought to a name. She replied that they had quite easily settled on Eugene. She said that it had been Gracie's suggestion and that they were all in accord. Dr. Abraham allowed as how that was a fine name, but thought that by giving the child a name they might find it hard to make him a state ward. After he spoke, his countenance became a bit more clouded. "And have you all," he asked emphasizing the word 'all,' "considered what it will mean to raise a Mongoloid child?"

Gemma replied, "We have talked, and we are certain that we can do it. As you yourself said, 'We can love him,' and we shall."

"You must know that it will not be easy. These children can be difficult at times, as I have seen over the years. Sometimes it is best for families to place them in an institution where they can be with their own kind — when the time is right. It is always hard at first, but families adjust. With your husband's work, I'm sure you will understand," Dr. Abraham said. Dr. Abraham could see that there were differences in this family, but he also knew that some families who had started out with good intentions could not manage the challenges attendant to raising an imbecile. There were sometimes just too many difficulties in the family, he said. And for that matter in society, too. Still, he was sympathetic to the Wellbournes, and tried to convey his support to Gemma, but he felt obligated to make certain of their commitment.

In the weeks that followed, word spread that the Wellbournes' child was borne an imbecile and that they were intending to keep him. By the time the gossip had churned its way to Ardis, which, owing to her connections throughout town, was not long, her Irish dander had been doubly aroused. To be sure, she was an ever-present buttress to the Wellbournes. It is more accurate to say that she was not only a stone wall that shielded them from the contempt now arising in the community, but that her wall came with parapets and cannon emplacements. In no way was she prepared to let them face an onslaught without a fiery response. And in that, she was just the kind of friend that the family needed. It was from her many kindnesses that Gemma, and not surprisingly Gracie, grew in their determination to raise baby Eugene the best way they knew how. Nigel was a bit slower to come around. Ambivalence plagued him for a time, it being so dastardly hard to reconcile the mission of his work with the bonds of family and his growing love of a sweet, innocent child. Harry Laughlin certainly did not make it easier, for he was quite emphatic about the need to separate defectives from society. He'd made that clear to Nigel on their first meeting. Moreover, he had begun to argue for a more proactive approach by using negative eugenics, the means for a more rapid elimination of the problem, and that was something that Nigel would never be able to fully accept.

CHAPTER VIII

Nigel adjusted to the reality of a new baby, if not yet to his disability. Although his sleep routine was modestly impacted at the outset, long hours at the laboratory insulated him from many of the imperatives of child rearing. Those fell to Gemma, and to some extent to Gracie, who despite her young age took on an ever-increasing share of minor household duties.

Gracie worked to develop Eugene's muscle tone and improve motor skills that come more naturally to normal children his age. Dr. Abraham recommended specific exercises, which Gracie dutifully carried out each day. As a consequence, Eugene physically developed at a rate that, in the doctor's judgment, approached that of other children. Given their time together, Gracie and Eugene began to develop an exceptional bond, and owing to their closeness, Eugene soon began to mirror her natural cheeriness. Others said that his good nature owed to innate tendencies. For Nigel, the origin of his disposition was a matter of persistent and rigorous contemplation, and was in his estimation worthy of future study. He initially considered the trait to have been one Eugene inherited from Gemma rather than himself, but that, too, was a matter that would have to wait for future consideration.

For most of Eugene's first year, Nigel was conflicted in ways that required him to resist the imperatives of his work, and balance philosophical disregard for the disabled with the inclination to care for, and even have affection for his son. At times, he could not suppress feelings of guilt for having biologically contributed to Eugene's affliction. Coupled with the stigma of rearing an enfeebled child, these feelings sometimes created an emotional wedge between father and son. Still, he had to admit that Eugene was the physical manifestation of his deep love for Gemma. That his wife and daughter so readily accepted Eugene was a constant reminder of his fatherly obligations. Despite Nigel's ambivalence, Gemma had none. Eugene was her child. He was a gift. He would be loved. He would be fine. Nothing else mattered.

Wellbourne's metamorphosis was not at first apparent. With time, however, he could not restrain affection for his infant son's beguiling personality. Even the most deeply flawed individual has some feelings of kinship. No one could deny love for a child who so readily offered joy, a child whose little face brightened so impishly on recognizing friend and family. That Eugene was dependent for his very existence made it all the more difficult to retain the coldly analytical detachment that might otherwise have been a eugenecist's predilection.

When Gracie began school, she saved each of her school lessons, so that she might at an appropriate time begin to teach them to her younger brother. Even with his developmental limitations, she believed that he would have the capacity to learn, if only at his own pace. And though there might be a limit to what he could in the end absorb intellectually, she would see to it that he would go as far as he could. Gracie's determination strengthened the family bonds, to which even the most hesitant of the Wellbournes succumbed. By the time of his first birthday, it was inconceivable to almost every member of the extended family, which included the Gilroys, that Eugene would ever be institutionalized.

Eugene lived within the confines of his home almost exclusively during his first year, a fact dictated more by the cycle of infant rearing than hesitancy on Gemma's part to bring him into the broader community. For that, Nigel had been grateful, because it was not long before word of his having fathered a feebleminded child spread through town, and thus among his coworkers. While nothing was spoken of it in his presence, he suspected from their furtive glances and stifled conversations that he and his family were the subject of sometimes disparaging remarks. Not the least of the tension associated with such gossip was Harry Laughlin's persistence in emphasizing the imperative of segregating defectives from society. It was a spoken refrain, and the subject of his numerous journal articles and speeches. Institutionalization spread throughout the nation as a result, and all of this burdened Nigel with an awareness of his own hypocrisy, not to mention a growing tension with Laughlin.

As he grew in strength and ability to function well among others, Gemma began taking Eugene along on visits to the Gilroys' boarding house, and later on her forays into town. Unswayed by social convention, Gemma felt no obligation to hide, or as a defiant gesture to flaunt her child. To her, he was to be socialized as any other child, subject to whatever constraints his own family chose to impose, regardless of social strictures. Consequently, in time it became for some a revelation that an imbecilic child could be accepted as any other child. This owed particularly to the Wellbourne family's simply going about their business like any other normal family. It wasn't that they were unaware of social judgment; they were simply unmoved by it. Of course Eugene's comportment in those early years was not easily distinguishable from others his age, despite a somewhat atypical appearance.

(What was it that Laughlin first said of those like him? '... they have a non-conforming appearance.' The fact is, the boy was cute. I will say it again. Eugene was an utterly beautiful child from the very start, and he brought a smile to all but the most fearful and hard of heart.)

[Yes, and I recall that you also had quite early grown attached to the child, as I myself did. But then, was he not fearfully and wonderfully made?]

By his second birthday, Eugene was a familiar sight in town. Of course, he was by then a regular at the boarding house, where there were many knees to bounce upon, and many a motherly bosom in which to nestle. In matters of play with Anthony and Gracie, Eugene was invariably the center of activity, mirth, and sometimes toddler mischief. Eugene evinced a budding sense of humor that disarmed even the most hesitant of those he encountered, despite his presumed limitations. Whether by mimicking strangers, or through playfulness, he easily charmed his way into people's hearts.

Let it not be said that he was in every way a perfect child. For he had a decidedly stubborn streak, which he exhibited when frustrated. It could arise from parental limits on some action he was about to take. Oh, and he was later to become a runner — and a fast one at that, but more shall be said on that at a later time. Running was a habit that, like stubbornness, suggested more than a little willfulness. Discouraged by Gemma, willfulness was deemed beneficial by Nigel, who saw in his son a characteristic that would enable some level of success in life. Nigel had noted some of the same attributes in himself, and though Eugene did not possess many of the other traits that marked his own or Gemma's personality, Nigel took comfort that at least some of his desirable characteristics had indeed been passed down from father to child. Perhaps this, too, should be a matter for future study, he thought.

By the end of Nigel's first two years at the ERO, he had been thoroughly immersed in compiling data from hundreds of field reports and questionnaires. Tabulating, calculating and coding information from the multi-page Record of Family Traits and Individual Analysis Cards was arduous and not particularly fulfilling. Naturally, Nigel longed to expand his role. From the outset, it had been a challenge to find willing survey subjects, besides those who had no say in the matter, like those then living in a growing number of state institutions for the insane and feeble-minded. Consequently, the lab resorted to mailing questionnaires to professors of biology, relatives of deceased luminaries gleaned from obituaries, and those listed in Who's Who. During that time, Nigel made enhancements to the various data collection forms, improvements that were duly noted by Davenport and Laughlin. In the same period, Nigel made several recruiting visits to Wellesley, Radcliffe and Pembroke, as the lab continued to add to the field force. All the while Nigel sought ways to improve field force selection through refinements to the profile against which prospective candidates were evaluated. It was difficult to find persons who had the interest, intellectual capacity, and tact, coupled with persistence, to probe sensitive matters that for families of defectives sometimes went back generations. Nigel observed that young women who were more attractive had greater success in getting family members to open up about degeneracy within their families. It was a trait that Nigel took pleasure in adding to his desired hiring profile.

Perhaps most challenging of all was finding young women who in all other respects seemed qualified, but who were also willing to relocate for a time and work in remote state hospitals. Nonetheless, Nigel began to develop an accurate feel for the necessary qualities. As a result, the field force was beginning to stabilize. This aspect of his work was particularly enjoyable, as it gave him the opportunity to socialize with many bright and attractive young students. In spite of mild social awkwardness, Nigel had become somewhat more at ease around young women. As was true with his experience at Pearson's lab in London, Nigel's competence was starting to be recognized, and thus his opinion was increasingly sought by Laughlin.

"Nigel, that new hire … what's her name … Lydia? She seems to be taking to the work quite well. Another good find!" offered Laughlin as he approached Nigel's desk in the workroom.

Looking up, "Oh, thank you, Harry," Nigel replied somewhat distractedly.

"Say, I don't know if you've seen this yet. It was only just published last year. It's from Goddard's work at the Vineland Training School," Harry said, handing him a book.

"Goddard, right. I'm familiar with him. Hmm … 'The *Kallikak Family, A Study in the Heredity of Feeblemindedness*.'" Nigel said, reading the title. "Have you read it?"

"Yes, and I think you might as well," Laughlin answered. "It's a good summation. More importantly, the book does what our statistical analysis can never do. It makes the subject much more personal. It tells a story, which I must say elicits a rather powerful emotional response. Kind of hits you in the gut. If we're ever going to have an impact on the problem, we have to move beyond just producing pages of statistics. Oh, those are necessary to support action … to provide an understanding about the scope of the problem. But to get anyone to take action, we need to also appeal to people's emotions. I'm doing a bit more writing on the matter myself now, thanks to you coming aboard," Laughlin added.

"Where will you be publishing your work now?" Nigel queried, more eagerly engaged in the topic of discussion. Even then, Nigel could not forget the slight he felt for Pearson's failure to grant attribution for his own research contribution.

"I'll be publishing a newsletter — *Eugenical News*. It'll summarize much of our research, and that of our associates around the world.

It will also tell some personal stories, which brings me to my point," he said.

"What's that?" Nigel inquired.

"You've been a great addition, and have freed me to do more of what Davenport has meant for me to accomplish; that is, preaching a better way, if you will. Our research work here is the foundation, but science must have a social purpose. Science must alter the way things are done if it is to be of service. For us to accomplish the societal changes that we know are sorely needed, we must first change the way society thinks. Now, back to my point. I know you have a keen interest in the subject of assessing and improving the germ line for intelligence. My point in giving you the book on the Kallikaks is so that you can gain some perspective on our good friend Goddard. Though his book touches on all manner of social inadequacy, he also addresses intelligence as a foundational element. As I'm sure you are well aware, he has recently translated into English the Binet-Simon Measuring Scale for Intelligence. I think you and I should pay a visit to Vineland to learn more about this work," Laughlin said, beginning to smile broadly.

"That would be splendid," responded Wellbourne. Noting Laughlin's change in expression, he added, "Is there something else?"

"Yes, I would like you to learn to administer the test. It would be a great help in our research, and I think you might also evaluate how and whether we could train our field staff to administer it as well. I think the effort might prove useful in accelerating our work. As you know, a few hospitals have begun to use the test. It's said that Goddard has already sold thousands of copies. Still, many hospitals lag behind. We might move things along by leading the effort. What do you say?" Laughlin said, now beaming with enthusiasm for his own gracious offer.

"That would be excellent. I certainly would welcome the opportunity to meet Goddard, and yes, to learn his test. I'll get busy on the book tonight."

"That's fine. I'll set up a meeting some time in the next few weeks and let you know. I assume you are okay to travel now that your son is a little older, which reminds me, how are Gemma and, ah …?"

"Eugene."

"That's right, Eugene. How are they doing? Pansy and I sure would like to get over and meet the little fellow," offered Laughlin.

"They're doing fine. Emm, that is, Gemma is well. Eugene has been a little off recently. Normal childhood sickness, I think. Nothing serious. He should be fine. Perhaps when he is feeling better, we can let you know a good time to visit, that is, between naps and feedings," Nigel said, chuckling.

"It's a busy time, I know. Perhaps when things are more settled. Anyway, I'll let you know when Goddard can meet. I'm certain you and he will find much to talk about," concluded Laughlin.

"I look forward to it. Now, I must get back to tallying. Oh, before you go, I will have that new outline of field-worker training to you by the end of the week. Should I pencil in a segment on IQ testing?"

"Yes, why don't you? That would be fine." replied Laughlin, turning to head to his office. Looking over his shoulder, Laughlin added, "Say, what do you make of what's going on in Europe?"

"More war, it seems. Typical for the Continent — the usual German expansionary aims. Fortunately, it's not our fight," Nigel concluded. Laughlin shrugged and went back into his office. Nigel returned to the task at hand, feeling more encouraged for the expansion of his role.

That evening, troubled by Laughlin's suggestion for a visit, Nigel tried to delicately broach the subject with Gemma. He feared she might be of a different mind on the matter. "Gemma, I hesitate to bring up a delicate subject just now, knowing that you have been so long deprived of a good night's sleep."

"Bring what up, dear?" she inquired. "I am fine. Yes, I have been awakened at night, but I have ample occasion to rest during the day. You needn't trouble yourself. Just what is it you would like to talk about?" Gemma asked.

"Harry once again suggested that he and Pansy would like to drop by. We've put this off for so long. I am afraid that I have exhausted all excuses. I know he suspects something. Certainly he has picked up on the gossip. It seems everyone else in town knows. I suppose he just wants to have a look for himself." Nigel grew more agitated.

"Well, I do not think any harm can come from a visit, and I am certain Pansy will be understanding. Ardis has told me so herself. Pansy has on occasion hinted privately to Ardis of her desire to have children. Perhaps it was not meant to be, but I do know that she cares for children, and there is no reason to believe she will feel differently about ours," asserted Gemma. She was beginning to show her own agitation at Nigel's intransigence on the matter. "I am wondering if it is you who has the problem. Perhaps you are ashamed of your own son," she said as tears welled up.

"I most certainly am not!" he answered. "I love our son. It's just … well, you know. Harry is adamant that all feebleminded children should be in an institution. You have heard him say so himself, '…for their own good, all the feebleminded must be provided custodial care, and where appropriate suitable training' … blah, blah, blah."

"Isn't that what your work is all about?" Gemma asked, now challenging Nigel to account.

"Yes, of course," he responded, trying to restore a more civil tone to the conversation. "At least for those who lack qualified parental oversight. Unlike most of the feebleminded I have encountered, we are able to provide such care. In any event, Eugene is far too young to even consider placing in an institution," Nigel added.

"I will not hear of it. If necessary, Gracie and I can provide whatever training he may require. As long as I am alive, our son will have a home with us. I do not care what the community or Harry Laughlin has to say about it, and as far as I'm concerned, you are more than welcome to invite the Laughlins to our home. In fact, I insist on it. If you do not, I myself shall!"

By then, it was apparent to the sometimes obtuse and hopelessly introspective Wellbourne that Gemma had once again gotten the better of him. No more was said of it until Gemma announced the date for the Laughlins' visit to their home for tea and cakes.

The Sunday before Nigel was to depart for the Training School at Vineland, Harry Laughlin and his wife Pansy stopped by for a scheduled afternoon visit. Nigel put on his most welcoming face when greeting them at the front door, but it was a strain. He was certain of Laughlin's judgment, and though he knew his superior would comport himself with conventional decorum, he also knew there were likely to be consequences later. He did not expect anything in particular from Pansy, only that in deference to Gemma she would outwardly display kindness and support.

"Please, do come in and make yourselves at home," offered Nigel. "We are delighted you could visit at last, now that things have settled down. Allow me to present my wife Gemma. This is our eldest, Gracie, who is six. Our newest, Eugene, is just waking from a nap and will be out shortly."

Pansy gushed, and Harry mumbled some pleasantry to no one in particular. Then, turning to Gemma, he said, "We are pleased that we could all be together at last. I feel we've come to know you well from what Nigel has told us. Please forgive us for not calling on you sooner. I … we know it has been a particularly busy time, what with adjusting to a new home, and having a baby and all." Turning then to Gracie, he said, "And Gracie, we likewise have heard so much about you starting school. Perhaps you recall when we first met. I'm afraid you caught me in a particularly trying moment, but then perhaps you have forgotten all about it. In any event, your father has spoken of you often, and we know he is so proud of you."

"Pleased to meet you again," said Gracie. She smiled kindly, and added, "I only partly remember our first visit, but I do remember getting to see a groundhog with Mr. Davenport." They all chuckled.

"And he no doubts remembers your visit as well. You must come by and see him again some time," Laughlin replied. "Say, where would you like us to set this cake that Pansy has baked?"

"You can just put it on the table there. If you will excuse me, I will go check on Eugene. He should be getting up from his nap about now," Gemma said. "I'll only be a moment. Perhaps he will be awake enough to greet you all, and then, if you would be kind enough to allow me time to feed him, I will rejoin you. In the meantime, can Nigel bring you a refreshment … tea perhaps?" said Gemma.

"That would be lovely, thank you," said Pansy. "Please don't hurry yourself on our account. We are happy to accommodate your schedule, or I should say the baby's schedule."

Gemma returned with Eugene, holding him in her arms, his head nestled tightly to her neck. She turned so that they could see his face, unsure what expression they would encounter. She feared that he might be cranky, or shy upon meeting strangers just after awakening. To her surprise and the surprise of her guests, after first glimpsing Gracie, Eugene was smiling broadly. Laughlin stood back from the others and observed carefully. From a distance, Eugene seemed small for his age. On closer examination, Laughlin could see that he resembled many of those whom he had observed in mental institutions. The startling recognition caused him to recoil. Laughlin was certain no one could detect his reaction. Pansy cooed, and smiled back at Eugene. Gracie walked closer, reached out and tickled her little brother, and he soon was laughing. Seizing the opportunity, Gemma did not hesitate to offer Eugene to Pansy, who took him in her arms with gratitude. He did not resist or frown, but continued to smile and laugh. As she swayed and spoke with him, it did not take long before Eugene was playing hide and seek with Pansy, alternately burying his face against her shoulder, then turning to look into her face with laughter. Laughlin continued to stand back and study the child. There was noticeable tension at the corner of his mouth and eyes. Nigel looked on pensively.

When Gemma excused herself to feed Eugene in the kitchen, Nigel cut the cake and offered a fresh pot of tea. The three adults chatted a bit, while Gemma fed Eugene, a process that was usually quite messy. She knew that his manner of eating would not be suitable for the Laughlins, at least while they were themselves enjoying a meal. Before long, Gemma returned to the dining table to join the group, but by this time Eugene had become cranky.

Sensing that it was a good time to depart, Laughlin suggested, "Pansy, dear, I think it's best that we go so that the Wellbournes can resume their routine." Rising, he said, "Thank you very much for the visit. I hope we can do it again sometime soon, perhaps next time at our house. It was a pleasure to see you again, Gemma ... and Gracie ... and to meet you too, young man." He nodded stiffly toward the child.

It had been obvious to everyone that Harry was not particularly relaxed during the course of the visit. Both Nigel and Gemma observed that he stood some distance away when Eugene was brought out to meet them, and remained standing a safe distance apart from the group while Eugene was in the room. Pansy, on the other hand, had been perfectly at ease during the short visit, and had taken an immediate liking to Gemma and the children.

(By this time, one has to wonder about Laughlin's relationship with children, not just Eugene, but any children. Perhaps his and Pansy's childlessness owes less to a biological artifact than his having been regularly visited by a succubus.)

[Egad, Counselor; that is harsh! You need to simmer down or find yourself meeting with some calamity.]

The following Tuesday, Gemma received a post from Pansy with an enclosed card on which she'd written a brief note thanking her for the visit. She remarked that it was such a joy to meet Gemma's children, and that she felt it a special gift to be able to hold Eugene. She added that she hoped that the two of them might get together soon.

Wednesday of that same week, Pansy stopped by for a brief, unexpected visit while on the way to the market. She apologized profusely for stopping by unannounced. Gemma assured her that visitors were always welcome. Her visit was welcomed so openly by Gemma that Pansy was moved to confess in the most apologetic way that Harry was not especially warm with children. She revealed that they had not been able to have any children themselves, and perhaps this was what accounted for Harry's discomfort during their earlier visit. Though unexpected, her confession was warmly received. Pansy said that she hoped Gemma was not put off by her awkward admissions, and that Harry really did appreciate the opportunity to visit them in their home. Despite the deeply personal revelation, the two had a delightful time becoming better acquainted. Gemma suggested that they meet again soon, it being so seldom of late for her to have lady friends stop by. Gemma suggested that Pansy join her at Ardis' boarding house the following Saturday, adding that she usually took Gracie and Eugene over to play with Maria's son Anthony each Saturday.

Over the ensuing weeks, these visits became part of each woman's regular routine, and Pansy Laughlin soon found herself, for the first time in the Cold Spring Harbor community, among real friends. All these goings-on made Harry Laughlin increasingly uncomfortable, though for a time he said nothing. Whatever verbal restraint he managed to muster owed to the fact that he was just pleased to see that his dear Pansy had recovered some outward joy in her life. He was beginning to understand that his work might have become a bit depressing to those around him.

CHAPTER IX

By the summer of 1916 Nigel Wellbourne had become proficient in the administration and scoring of the Binet-Simon intelligence test. Goddard imbued Wellbourne with an unflagging commitment to IQ testing as a way of sorting persons with varying degrees of feeblemindedness. Admittedly, there were a relative few inmates that demonstrated general capability in the upper scoring ranges. The exceptional inmate might have performed well on only one or two subtests, but the overwhelming majority scored well below the norm, especially those who had limited English language ability. The intelligence scale proved quite useful in making inmate assignments. The high-grade defectives could be provided a few years of basic schooling before being assigned to vocational training. The lower grade defectives could only be given routine pick-and-shovel work or housekeeping duties, while the more seriously impaired required only basic caretaking.

Unsurprisingly, Nigel embraced IQ testing with such fervency that he became one of Goddard's most loyal disciples. His loyalty brought reciprocity from Goddard, who later introduced Wellbourne to two other pioneers in the movement. Both had been corresponding with Goddard about expanding the use of mental testing. As with testing the feebleminded, they asserted that intelligence testing could be more broadly useful in determining education and career paths for normal persons.

Even as Nigel became adept at administering the test, he proved even more skilled at selecting field workers whom he could entrust to administer the test. One young field worker recruit in particular caught his eye. Their first meeting was to mark the beginning of a dramatic turn in his life.

Sister Wellesley, fairest of the Seven Sisters, was once again to yield her fruit. This time it would be in the form of a beauty unadorned, a veritable goddess, a mathematical prodigy, and soon to be alumna. Julia Benson reminded Nigel in so many ways of his beloved Gemma. She was, absent embellishment, classically radiant, graceful in form, and in every circumstance exceptionally self-possessed. She would not be thrown off by the disagreeable or untoward picture that Nigel had vividly painted of the work, which he cautioned involved picking and sorting from among the detritus of society. Occasional unpleasantness was, in her words, a simple fact of life, and thus required of an advancing society that it be willing to rid itself of untidiness. She had, he thought at her turn of phrase, easily absorbed the nuances of the movement's language, and demonstrated fluency in its use. In so many ways she was superior to those whom Laughlin previously engaged in the work.

While not unfeeling, she could, like Nigel, hold sentiment in check in order to pursue the imperatives of scientific inquiry. During her interview, she offered that her primary interest was in science and improving the lot of man. Her evident emotional detachment seemed to make her particularly well suited to the job, especially given her other qualities.

Upon confirming her willingness to be temporarily posted somewhere in the Midwest, she was offered employment as a field-worker, pending her imminent graduation. Nigel told her that along with other newly hired field-workers, her training would commence at the Cold Spring Harbor laboratory in mid-June. Thereafter, it was expected that she would be assigned to the Kansas Home for the Feebleminded sometime in mid to late summer. If things worked out, he said that she would also have the privilege of traveling to her first post in the company of both himself and Mr. Harry Laughlin. Since Laughlin had unspecified business with the institution director, Nigel would help her begin training the institution staff to implement IQ testing. Once the training was underway, Nigel indicated that he and Laughlin would take a brief excursion to Hutchinson for the Kansas State Free Fair. There they'd attend a Better Baby Contest, which was sponsored by the Kansas Eugenics Society. On their return to the institution, they would resolve any outstanding issues and then say their goodbyes.

Nigel returned from the recruiting trip to Wellesley ebullient at the consummation of his successful hire. The spring weather was exceptionally pleasant. The flowers which Gemma had planted in the window boxes and along the front of their cottage were in full display. The summer evening was redolent of new life, and with daylight sufficiently extended, he eagerly anticipated the enjoyments of a pleasant family evening. Hopping off his bicycle at the end of the sidewalk leading up to the house, he paused to observe Eugene playing alone in front of the house. By his fourth year, Eugene had matured sufficiently that he could attain some level of independence in outside play. Kneeling and bent forward at the waist, he appeared to be blowing gently on an object that was presently out of Nigel's view. On closer observation, he seemed to be holding something gently in his hands. Curious, yet not wanting to disturb the child, Nigel held off calling out to his son.

At the side of the house, Gracie was likewise engaged in play, though more actively than Eugene. Nigel could not hear her, but could see her lips moving. She appeared to be conversing animatedly with her dolls. Periodically, she would glance over at Eugene. Neither was she aware of Nigel's presence, whose outline was partially obscured by shrubbery.

Turning his gaze back to Eugene, Nigel was taken by surprise when a small bird, a sparrow perhaps, fluttered out of the boy's cupped hands. Turning to follow the flight of the bird, Eugene spotted his father and called out, "I did it! I saved him! I saved a bird!" He clapped his hands close to his chest as he spoke. It was the manner in which he had, since a very young age, expressed pleasure at his own accomplishments. Though his speech was halting and mildly thick-tongued, Nigel could understand, for Nigel heard the same manner of speech in his many state hospital visits. It was simply a byproduct of mental and physical affliction. To some, such cumbersome speech was off-putting, a mark of imbecility. To the people who had by then grown to know and accept Eugene, it was endearing, and to those who opened their hearts, Eugene remained every bit the adorable child he had been in infancy.

Spotting his father, Eugene raced down the walk and hugged him around the legs, then distractedly ran off to see what his sister was doing in the side yard. Nigel wheeled his bike to the front door, and entered to find Gemma in the kitchen preparing the evening meal. "You won't believe what I just saw. Eugene was kneeling by the front garden, holding something in his hands. A bird fluttered out, and he announced, 'I saved it.'"

"Where was Gracie?" Gemma asked. "She was supposed to be watching him."

"She was …" Nigel answered, sounding somewhat defensive. "… she was in the side yard playing, but she kept an eye on him," he responded.

"Something hit the window earlier. I suppose it was a bird. Eugene must have found it. Sometimes they are just stunned, then wake up and fly away," she said.

"I suppose, but we should let him think he saved it, don't you agree?" Nigel said.

"Of course, Nigel. It's good he should learn that he can care for living things," Gemma concluded. She dried her hands and walked over to greet Nigel with a kiss.

"Gemma, darling, what's for supper?" he asked. "I'm famished."

"Pot roast. It should be ready in half an hour. How was your day?"

"Quite good. Oh, I should alert you. I will be traveling a bit more this summer," he said.

"Where to this time?" Gemma asked.

"That new field-worker we just hired should be ready for placement then. She'll be going to Kansas … Winfield, Kansas. There's a hospital there. She and I will be training the staff in testing. Harry will travel with us. He has a meeting with the director."

"How long do you expect to be gone?" she asked, now sounding a bit impatient.

"It should only be about two weeks. I can send a telegram from there when we are ready to leave for home. We'll also be attending the Kansas Free Fair in Hutchinson. Seems they've made quite a smash with their Better Baby Contests. They're all over the Midwest now.

Harry thinks we should see it. He says he has some ideas on how we can make it better," Nigel gushed.

"Better Baby Contest, eh? Better than whom … or is it what?" she added trying not to sound caustic. "I think we have two blue ribbon winners right here," she added trying to change the tone of her comment.

"Listen, Gemma. I'm concerned about being gone so long. Will you and the children be all right in my absence?" he asked.

"Yes, dear. We'll be fine. Ardis and Cork pick me up once a week for shopping, and Cork is always willing to drive us wherever we need to go in an emergency. Gracie can fetch him if I'm occupied with Eugene, or we can run over and get him ourselves if we have to. That reminds me, we got a letter today from the Wellstones. It seems their son has gone off to war in France. They ask that we remember him in our prayers." Gemma could not hide her feelings in sharing this news. The prospect of war and loss, for that matter loss under any circumstance, was troubling to her. Nigel saw it and walked over to comfort her.

"He'll be fine," Nigel said, trying to buck her up. "Nothing more gallant than a strong British lad, especially a young Wellstone," he noted, still hugging her.

To change the subject, Gemma asked, "How are things going with Harry these days?" She knew that even under the best circumstances Laughlin's opinions could border on stridency, a trait that had often caused strain on Nigel.

"We're getting on fine for the most part. He is preoccupied with propounding the benefits of eugenics and has given me sufficient latitude to do my job. That said, he has become a bit of a stodge on the matter of segregation of the socially unfit. He says it often enough in my presence as to suggest it is time that Eugene be sent away. That has begun to rankle a bit. Still, he has never mentioned Eugene by name, and even occasionally inquires about him. He's on another mission, at least for now."

"Oh, what about?" Gemma asked.

"Many think he's not altogether wrong on this one. The number and kind of immigrants, or should I say the quality, he says has begun to degrade our prospects as a nation. I fear he has crossed a line on this one, but have chosen to remain silent on the issue. Anyway, he has fashioned a program of research into the matter. He thinks that we can use the intelligence scale to make the case about limiting immigration to those who are best suited. Means more opportunity for me, I suppose."

"Perhaps," Gemma noted with a broad smile, "it is well that we came when we did." Nigel found her mild quip to be amusing, and was distracted from delivering yet another tedious lecture to Gemma.

When the day of departure arrived, Cork and the others picked up Nigel at his house. He kissed Gemma at the door, and bade the children goodbye, barely containing his excitement at the prospect of venturing further into the heart of America. Admittedly, his excitement owed in part to leaving the routine of home life, which by then had settled into a somewhat dull pattern, in spite of the occasional challenges of raising a not infrequently recalcitrant Mongoloid child. More than that, though, it was the excitement of traveling with a man who was partly responsible for what by then had become a global movement of human betterment. Additionally, he would be in the company of a bright and lovely young woman, one who seemed to hang on his every word.

Arriving at the New York station, the three boarded the 20th Century Limited, which would take them as far as Chicago. Laughlin had planned for them to spend the night at the Palmer House, where they would have dinner with Professor Edward C. Hayes of Illinois University. Hayes, a sociologist and supporter of the eugenics movement, had just the previous year published a highly regarded textbook on sociology in which he advocated the use of eugenics to improve birth rate selectivity. Laughlin had asked to meet with the professor to discuss the prospect of making a classroom assignment for Hayes' students, which would require them to complete family history questionnaires.

The three would leave the next day for the remainder of their trip. All totaled, it would take nearly forty-eight hours to reach Kansas City, and after some delay, another few hours travel by train to reach Wichita. There, they would catch a bus to Winfield. When they finished their business at the state hospital in Winfield, Laughlin and Wellbourne would take a bus to Hutchinson for the Kansas State Fair.

Laughlin had reserved three seats in the passenger car so they could be together for the bulk of the trip. He apologetically revealed that he had also reserved a roomette for his own use. That was, he advised, so he could work late into the night on what he said were overdue budgetary reports. In this way, he went on, he would not disturb them while they rested in coach. Still fearing that his rationale for a private room was not warmly received, he explained that given the duration of the trip, its associated expense, and the fact that they were traveling in mixed company, it would be best to have only one small room. Laughlin did however spend nearly all of the daylight hours in the company of his traveling companions. This, he reasoned, would afford him the opportunity to get to know Julia better, and they all could discuss their respective work plans.

Harry sat opposite Julia, who occupied a position nearest the window. Nigel sat next to her. Harry had suggested this seating arrangement as it afforded him a better opportunity to have a free and open conversation with both of them, and because it gave Julia the preferred seat next to a window. The opposing seats were sufficiently close together that the occasional touching of extremities was unavoidable, and for Harry not particularly unwelcome. As their seats were located at the end of the rail car next to a bulkhead, they were able to converse without fear of being overheard, an arrangement necessitated by the nature of their work.

While the other two got better acquainted, Nigel was fully absorbed in the most recent issue of *Eugenical News,* Laughlin's house organ, which was designed to unify the many geographically dispersed employees and affiliates. Nigel first turned to the summary of field reports, an article derived from his monthly activity report. By field-worker name, it summarized the tally of sheets, charts and persons surveyed. This regular column was to demonstrate for readers, particularly the lab's benefactors, the quantity of work completed, as well as to provide motivation for the field staff. Nigel then read Laughlin's summary of a manuscript entitled "Being Well Born," in which the author, Dr. Michael Guyer, said that 'the germ of self-control was not present in some cacogenic families.' It was, Nigel ruminated, a phrase that was both apt and memorable, when he overheard a bit of Laughlin's conversation with Julia.

"… voice of eugenics cries out for better parents," Laughlin said. He went on, "The lowest 10% of human stock are so meagerly endowed by nature that their perpetuation constitutes a social menace. And certainly the enfeebled … all of the feebleminded must be institutionalized." Julia listened attentively as Laughlin fixed his gaze on her. She suspected that he might also have been intending his last remark for the benefit of Nigel, as he looked once or twice out of the corner of his eye toward him.

Ignoring his last comment, she asked, "I understand the imperative, but how is this to be achieved when marriage is too often based on romance or physical attraction rather than eugenical suitability?"

"We have collected sufficient data by now to demonstrate beyond doubt that defective germ plasm is at the heart of social dysfunction. There can be no question. It is a matter of settled science.

(*There's that phrase again!*)

[**It does engender a certain amount of skepticism, doesn't it? Is science ever completely settled?**]

"And though we have until now primarily focused on educating society as to the benefits of eugenics, it has become quite clear that we cannot, and must not rely on informational efforts alone. To be frank, we must be more proactive, even forceful," he said. As always, Laughlin lengthened his spine, thrust his chin forward and swelled his chest when he spoke of taking action. Unable to pretend any longer that Laughlin was speaking only to Julia, Wellbourne glanced over at Laughlin, whom he assumed must have been seeking affirmation. Was it just one more of Laughlin's loyalty tests, Nigel wondered?

"I should say that our field work is foundational. Our data serve as proof as well as a call to action," Wellbourne offered, knowing that Ms. Benson should be given some validation for the more mundane work she would be performing.

"Dr. Laughlin, just what actions might we take to accelerate progress in this area?" Julia asked, hoping to come across as deferential.

"This is a very good question," he answered. Then turning to Wellbourne, "I can see that you have selected well, Nigel. Ms. Benson demonstrates a bias to action that I think will serve us well." Turning back to answer her, he said, "There are three main thrusts. I know that Wellbourne here knows them well. Of course, we must continue to build on our knowledge of inherited characteristics. That is the work of our fine field and laboratory staffs ... persons such as yourself. But the thrust of that work is, in the end, to segregate from birth those who are defective, to restrict the importation of the inadequate, and to limit the fecundity of those who would continue to produce inferior offspring."

It was a statement made with obvious finality, delivered in a manner that would seem to brook nothing but assent. The result was a terminal silence, at least for an embarrassing interval. Not inattentive to the effect his statement had on his traveling companions, Laughlin softened the rhetorical excess through conversational redirection. Inquiring of his mates whether either had ever traveled to Chicago, he offered his observations on the meal they could expect later the next day at the Palmer House. After a brief reply, Nigel returned to reading the *Eugenical News*, and Ms. Benson to looking out the window.

With due allowance for his seat mates to retreat to their own thoughts, Laughlin finally turned to Nigel and asked quietly, "How is everyone at home?"

"Everyone is fine. Thank you for asking," he replied. Not wishing to seem guarded, he continued, "Gemma is well. She has adapted quite swimmingly to her new surroundings. She's made several friends. Gracie is in school, and doing quite well."

"And how is everyone else adapting?" Laughlin continued.

"Do you mean Eugene?" he answered, doing his best not to sound defensive. "He is quite well. He is in many ways a clever boy, and keeps us on our toes. Too, he can be quite endearing. At least to the people who know him well," he added.

"I gather his limitations have not as yet presented problems for you, then," Laughlin went on.

"He is in many respects a normal child. Though I will admit he will not likely profit from schooling in the same way as other children … when the time comes," Nigel answered. He tried to avoid sounding defensive. "Gracie has already begun to share some early lessons with him. He delights in the time with her, and in some ways surprises us with what he is learning."

"Ah, yes … when the time comes," Laughlin sighed, then glanced over at Julia, who was reading a book and appeared not to be taking in the conversation. "And when the time comes, as it invariably does, do you imagine that you and Gemma will …" he paused, "consider placing him in an institution, where he might join with others of his … his circumstance?"

Nigel stiffened, then shifted in his seat momentarily before answering. Julia looked over at Nigel, then returned to the page she had been reading. Nigel settled himself, and replied, "We have made no decisions about Eugene's future. That said, he has been, and we expect him to continue to be, a lovely child. For that we are grateful. Thank you for your interest." Nigel's change in tone made it clear to Laughlin that he should not continue the conversation. At a time when he concluded that no inference could be drawn from his exit, he excused himself, and adjourned to his cabin, where he reminded them that he had budgetary work to complete. He reassured them that he would be joining them later for dinner.

"Your child, Eugene, is he …?," Julia started.

"Is a Mongoloid. Laughlin believes he should be in an institution. That is what we are all about, isn't it?" Nigel answered. "Our daughter Gracie is a normal child, if you might have wondered. In fact, she is quite advanced. We are a normal family, and have made no plans to remove our son." By then Nigel's anger had risen.

Julia touched his arm, and said, "I can see that you care for your son ... love your family. They must all be exceptional. They have a wonderful husband and father." She looked a bit too long in his eyes, then turned away. Looking out the window, she sighed, then said, "It must be hard ... this work and all."

"Julia, you should know that I am committed to the work. I have been for years. If we can ... if society can find some way to improve the lot of those who are afflicted ... if we could eliminate the suffering, why wouldn't we? Look, I have traveled throughout England, and many parts of the eastern U.S. I've seen the suffering. And I've counted the social and economic cost. For many years, the problem was hidden. We were once as a nation small and rural, so that the problems could not be readily seen. Now, we are living close together in large cities. Nothing is hidden from our eyes. Our economy is more industrial than agricultural today. Society needs highly functioning people. Why would it be so wrong to want to improve the lot of people?" By this time, Nigel was wrestling with his own thoughts, his own contradictions, and was not so much speaking with Julia as to himself. Regaining control, in the quiet interlude he added, "We must do what is best for our child. We must not confuse taking action toward people in general with doing things that are best considering individual circumstances. As a consequence, I must confess that Harry's dogged insistence on institutionalizing Eugene has become quite grating ... a cacophonous ringing in my head." When he finished, he was certain that he must have sounded hypocritical to Julia.

She turned back from the window and studied Nigel's face before continuing. "Mr. Wellbourne, I think you are a good man, and I know you always do what is right." At that she slumped lower in her seat and leaned against the window.

They sat together in silence. The lights of the rail car had been dimmed, and before long, she fell asleep. Nigel, too, had begun to nod off intermittently. As a conductor passed by, his head snapped upward, and he noticed that Julia's head rested tenderly against his shoulder. He hesitated for a time, leaned into her unobtrusively, then he, too, slumped lower in the seat and began to drift off.

CHAPTER X

Harry Laughlin had an agenda.

(Harry's life was an agenda!)

[Can you please just stop and listen for once?]

The first item on his agenda was for Laughlin's whole contingent to meet with the state hospital superintendent, Dr. F. C. Case, and his deputy superintendent. After introductions were made, Laughlin laid out the general goals for the visit. Dr. Case provided a brief history and overview of the organizational structure. With an abundance of pride, Dr. Case said that the institution cared for 589 persons on grounds that encompassed several dozen acres. The Kansas State Home for the Feebleminded included a farm, and several large multi-story buildings that housed inmates in separate dormitories for boys and girls.

The farm and its associated buildings, as with so many other similar institutions, enabled the home to be mostly self-sufficient, and also provided much unskilled work opportunity for the inmates. Unique to the Kansas institution were several producing oil wells on the property, which further helped offset operating costs.

Dr. Case pointed out that the institution's school offered a modified curriculum that encompassed at most a year or two of primary school. This was, he noted, simply a function of the residents' limited capacity for learning. It was judged that most could not profit from schooling, so instead received instruction in daily living skills. The girls who were deemed able received training in basket weaving, sewing, and related domestic labors. The boys were trained to do farm and general manual labor work. The rest were nominally cared for in large day rooms.

Following a quick tour of the buildings and grounds, the group gathered for a working lunch. Joining them was the medical director, Dr. Minshaw, and the chief steward, a Mrs. Nellie Peeble. During lunch, Laughlin outlined a proposed agenda for the week, followed by a detailed schedule for staff training. The training, which had been expressly requested by Dr. Case, was to be an abbreviated version of the ERO field-worker training. The intent was to give the institution's staff a basic understanding of principles of eugenics, as well as the ERO's specific data collection process. The staff was then to access inmate records, describe known inmate limitations, and arrange family member interviews. The staff would also receive an overview of intelligence testing, which Miss Benson would later be carrying out on selected inmates. Several staff members would be identified who, subject to further training, would continue intelligence testing. Once training was completed, Mr. Wellbourne and Miss Benson were to complete several questionnaires while the staff observed. After Laughlin finished his presentation, Dr. Case concluded the meeting by announcing that Miss Benson would continue her work at the home for up to a year, and that full cooperation with her was to be ensured. The meeting was adjourned shortly after 2 P.M., and the ERO entourage left for town to check in to a hotel. Their early afternoon departure from the institution was planned so that Julia could secure suitable housing for the remainder of her service at the hospital.

Over the next week, Nigel and Julia completed their staff training. Meanwhile, Laughlin met privately with Doctors Case and Minshaw several times. Laughlin had been well aware that the Kansas

Home for the Feeble Minded once had a reputation for vigorously advocating the sterilization of inmates. Sterilization, though not then without controversy, had begun in earnest well before Dr. Case's appointment as superintendent, but had tailed off. Dr. J. Barnes Tilman, Case's predecessor, was said to have been particularly aggressive in his pursuit of sterilization. During his tenure, he performed some 58 castrations and 158 sterilizations on what was then a small but growing population of inmates. He had concluded that sterilization, and castrations in particular, would reduce aggressiveness, self-gratification, and lust. In 1913, the legislature formally approved sterilization, though by the time of Laughlin's visit it was little practiced. Perhaps its restraint owed more to a cumbersome legal review process than any reluctance to perform the procedure. In 1917, however, a bill was introduced that simplified oversight of the practice, this time requiring only the approval of a state board of examiners rather than the courts. Thereafter, the rate of sterilizations began to pick up. Although Dr. Case kept meticulous health and hospital records, and reported biennially to the State Board of Control, no official statistics of sterilizations were ever included in his reports. It was in fact Kansas' earlier experience with sterilizations that had been of particular interest to Laughlin, and that was primarily why he had chosen to accompany Wellbourne and Benson on the trip in the first place. He hoped that he might learn enough from the Kansas experience that other states could be induced to accelerate their own practice of the procedure. Although it would take a number of years, he would indeed learn enough from this visit and his own research that he was later able to craft model legislation in 1922 which he hoped would be adopted by other states. Laughlin's model law would indeed eventually give rise to the practice being adopted in 37 states and a number of foreign countries, including Germany.

◆

Julia sat at a small table in a private office adjoining one of the day rooms. Nigel sat behind her and slightly to one side. In a box on the floor next to the table were the supplies she would need to administer the Binet-Simon intelligence test. The charge nurse brought in the first person to be tested.

"This is Brenda. Brenda is nine years old. Sit up straight and say good morning to Miss Benson, Brenda," the nurse admonished as she settled the child in a chair opposite Miss Benson. Brenda mumbled something inaudible. She rocked, and looked at her hands as she rubbed them nervously against her thighs. "Miss Benson will ask you some questions. I'll come back in just a little bit." Brenda did not look up. The nurse's starched uniform swished as the door clicked shut with determined firmness.

"Good morning," Julia said in her most soothing voice. "We're here to ask you a few questions so that we can get to know you better. Please be comfortable. There is no need to worry about anything. This will only take a few minutes, and should be fun. Do you have any questions before we begin?" Brenda rocked in her chair and said nothing. "So to begin, I'd like to first ask you your name," said Julia. "And can you spell it for me?"

"Brenda," the child mumbled while still looking down at her hands. She did not offer to spell her name.

"Good, Brenda. It is so nice to meet you," replied Julia. She was doing her best to put the child at ease. Nigel looked at Julia and nodded to reassure her that she was proceeding well. "Now, Brenda, before we begin, I'd like you to give me your biggest smile. I can see you are a very pretty girl, and I'd just like to have a good look at your face."

This seemed to relax the child. She looked up at Julia from under her straight bangs, and gave a shy smile. Julia's pleasing appearance and broad smile were reassuring. While Julia distracted her with a soothing voice and friendly expression, she inconspicuously placed the first test item on the table. It was a simple line drawing of a little girl with a running dog at her feet.

"Now, Brenda," she began, "I have a picture for you to look at. Can you see it clearly?" She asked. Brenda nodded. "Something is missing from the picture. Can you tell me what it is? What is missing from the picture? Please take your time, and look at the picture carefully. Then, tell me what is missing."

Brenda leaned in and looked closely. She hesitated for a second, then said, "The mother."

"Thank you," said Julia, smiling, as she made a note that Brenda failed to identify the girl's missing pigtail. Brenda smiled back at Julia, who removed the picture from the table. Julia knew that normal children of seven nearly always correctly identify the missing element from these simple pictures. She selected instead a test item that five-year-old children nearly always get correct. In this way, Julia would try to bracket the appropriate test level for the child. She placed a blank sheet of paper on the table, and next to it a sharpened pencil. Next to the blank paper, she placed a simple line drawing of a regular pentagon.

"This time, Brenda, I'd like you to use the pencil and the blank sheet of paper to copy the drawing. Do you understand?"

Brenda nodded.

"Go ahead, then, Brenda, pick up the pencil and copy the drawing on the blank sheet of paper I placed in front of you." Julia gave no further instructions as she waited for Brenda to begin.

It took a bit longer than normal for Brenda to clumsily pick up the pencil. She clutched the pencil in her fist, and carefully studied the image of the pentagon. Then she began to copy it on the same paper that had the original drawing rather than on the blank sheet of paper.

"Brenda, please use the blank sheet of paper for your drawing," Julia said as she moved the paper with the pentagon above the blank sheet of paper, and slightly out of reach of the child. She then pointed to the blank sheet and said, "You can draw it here."

Brenda made an attempt to copy the drawing, but hers had too many sides of varying length. None of the sides was straight, and she did not completely close the shape.

"Thank you, Brenda," Julia said with a bit more enthusiasm than before. She could see that Brenda was a little frustrated with her drawing, which seemed to make her more anxious. "Let me see that pretty smile again," she added, hoping to give Brenda a boost of confidence. Again, Nigel made eye contact with Julia, and nodded. Julia then said, "I'm going to read two short sentences. I'd like you to repeat them exactly as I read them. Do you understand?" she asked.

Again, Brenda nodded, but this time she looked at Julia and gave a shy smile.

Picking up the paper and reading slowly, Julia said, "His name is John. He is a very good boy." Pausing for a brief second for Brenda to process what she heard, Julia asked, "Can you repeat that please?"

"John is a good boy," she replied haltingly, but this time smiling broadly, confident that she had got it right.

It was clear to Julia that Brenda was struggling to respond to items that were normally completed by a five-year-old. To be certain of her assessment, she did one more item at that level. Putting out some coins, she asked Brenda to count out five pennies. Brenda counted three coins of mixed denomination, so Julia began again with items suitable for a four-year-old. Naming simple objects, repeating three numbers accurately, and correctly identifying her own gender confirmed for Julia that Brenda had the mental age of someone who was four years old. She turned to Nigel, and whispered her conclusion, and after making a note in the child's file, showed Nigel her notation that even the most basic level of schooling was not indicated for this child. When the nurse arrived with a second student, she asked, "And how did our Brenda do?" Brenda once again began to rock and rub her thighs.

Julia replied, "She did just fine. I've made a note in the file for you, and have added a comment, which you can read later." Turning to Brenda, she said, "Didn't you do well, Brenda? You did just fine." Brenda nodded and smiled broadly in reply, then shuffled out the door as the nurse placed her hand on the child's shoulder to guide the way.

Before beginning the process again, Julia turned to Nigel and whispered so as not to be overheard by the new arrival, "Such an adorable child. It's a shame …"

Nigel interrupted, "Yes, it's a shame we can't do more for such as these. That is why we do this work … to prevent such problems in the first instance." Julia nodded and returned to the task at hand.

After a long day of testing, Nigel debriefed Julia on the day. He remarked with the greatest enthusiasm that she had done a masterful job, and was most certainly capable of proceeding on her own. He added that, during her ERO training at the beginning of the summer, she had already demonstrated a level of skill and maturity beyond those of more experienced hires, and could easily be trusted to handle all aspects of her job. Nigel concluded their discussion by asking, "Julia, have you found your housing to be suitable?"

"Oh my, yes!" she replied enthusiastically. "The boarding house where I'm staying is quite comfortable, and the caretaker has been most accommodating. Plus I've met a delightful young woman who also boards there. She's about the same age as I am, and clerks at the bank here in town. I think we could become friends. So yes, I think I'll adjust to living out here just fine."

"That is excellent," Nigel said, holding the door for her. " Just know that you won't be out here forever. Say, I wonder ..." he paused, then continued, "as Harry will be dining with Dr. Case tonight, would you be free to have dinner with me?"

"Why, thank you, Mr. Wellbourne," she replied. "I'd be delighted." She curtsied slightly, then giggled.

"Wonderful," he said. "Since this is our last night in town — as you know, we're leaving for Hutchinson tomorrow — this will be our last time together until we pass through on our way home. Before we go, I'm very interested in hearing how this whole experience has gone for you. Shall we say 7:00?" he finished.

"That will be fine. I'll see you at the boarding house at 7:00, then," she answered.

Nigel could not help to notice that she blushed as she said it.

◆

Nigel and Harry discovered that the Kansas State Free Fair was not, in fact, free. The revelation was a great source of amusement to them for some time after arrival, and a matter of some joking later upon their return to Cold Spring Harbor. Admission was two bits. Two bits! Hog dogs, of which they each consumed two, cost a nickel each! The beers to wash them down were a dime apiece! Free indeed!

Avoiding the arcade, they headed through the barns that housed domesticated livestock, stopping only briefly by the horse barn, where Laughlin declaimed knowledgeably on points of equine conformation, husbandry, and breeding. As a longtime member of the American Breeders Association, and having formerly researched and written extensively on horse breeding in particular, his observations were well founded, though tedious. Without trying to appear obsequious, Nigel gave him his full attention.

The air was prairie hot and languid. The smell of cotton candy blended with manure and stale beer. Throngs of overall-clad farmers with wives in print dresses shuffled along aimlessly amidst a cloud of choking dust. Like locusts, the fairgoers consumed everything in their path, stopping in clusters to eat and watch, watch and eat, then drift on to the next point of interest. Unlike the locals, Harry and Nigel were on a mission; they needed to get to the Eugenics Pavilion before the last round of judging was completed. They had been directed to the site immediately on arriving at the fair, and were told to go just past the swine barn and take a left in the direction of the freak show.

"First fair?" asked Harry.

"First of this kind," Nigel replied. "I've been to a British fair, one that featured sheep herding and a bit of Morris dancing, but it was nothing like this," he said.

"How so?" Harry inquired.

"Without being prejudicial, I would suggest that this one is a good bit less refined. Though I must add that while Morris dancing is considered by some to be splendid, I think it every bit as peculiar as, say, paying money to observe persons whose physical afflictions make them an object for display." Nigel nodded in the direction of the entrance to the freak show. "Can you imagine grown men dancing with bells on their feet and flowers in their hats? That in itself is quite the spectacle!"

"Point well taken," Laughlin said. "Ah, there it is over there. I think we've found our spot," he said gesturing to a small building with a covered front porch.

They made their way through the benches arrayed in front of the building. The sign over the entrance said Eugenics and Health Exhibit. Nailed to nearly every available space on the front of the building were hand-lettered posters presenting various dysgenic phenomena, things like polydactylism, epilepsy, drunkenness and nomadism. A table was placed on the small stage that jutted out from the front porch. A child of about two, in diapers, was seated on the table. The child's mother stood to one side and looked on anxiously as her child reached out to his mother for succor. A nurse and a doctor examined the child, first with calipers, then with a tape measure and scale, as if examining a 4-H-er's prized piglet. When their work was finished, the mother scooped up her child and retreated to the inside of the small house, where she could be seen changing his diaper. Outside, the small gathering of mostly mothers waited excitedly for the announcement of the winner. A woman stepped out from the building and announced that the second annual prize for the best baby would be announced shortly, while the doctor and judges huddled to go over the day's results. A two-foot-tall loving cup was placed on the table by a helper, and a small poster was displayed showing the judging criteria, which mostly described attributes of physiognomy and temperament. Nigel and Harry sat patiently on a back bench, arms folded. When the results were announced, flashbulbs popped, children wailed, and reporters scurried to collect names. The small gathering of contestants surged forward to congratulate the winner, their own babies firmly in their grip. After the crowd dispersed, Nigel and Harry entered the small building to introduce themselves.

"Miss Pinchot, my name is Harry Laughlin, and this is my associate, Nigel Wellbourne," he said. "As I wrote to you earlier, we're from the Eugenics Records Office of the Carnegie Institution of Washington. We've come all the way from New York to observe your contest."

Miss Pinchot, head of the local chapter of the Eugenics Society of America, responded, "Well, we are most certainly delighted that you have come all this way. I must say, I'm totally thrilled to finally meet the head of the Records Office," she gushed. "I get the *Eugenics News*, you know. I'm quite familiar with your work."

"Yes, I am pleased to meet you as well. Nigel, here, works with me, and helps me write the news. He's in charge of our field force. In fact, we were just over in Winfield getting a new worker started at the school for the feebleminded."

"You don't say," she responded. "They certainly have their hands full over there — but they do good work, I'm told. It's a shame that we can't do more to prevent such problems."

"Indeed," replied Harry. "In part, that is why we have come out to Kansas. You see, while we think contests like yours are a good start, we believe there is more that can be done. These Better Baby contests, as you call them, do a good job of education on matters of public health — better nutrition, stimulation, hygiene, and such. But even the most beautiful child could, in fact, turn out to be an epileptic in two years."

"That is true, I suppose," Miss Pinchot responded. "Just what did you have in mind?" she added.

"Well, it seems to me that if we are to improve the stock, we have to delve into the whole pedigree, if you take my meaning. Let me be more explicit. It is the family that is at issue. We must encourage the development of better families, and that must be through the encouragement of more eugenical marriage. In addition, we must encourage those who come from a long line of the most fit to have larger families, while at the same time discouraging those less well adapted to the demands of modern society from having so many children."

"Fitter families is what we're saying," interjected Nigel.

Harry turned and stared at Nigel.

"Fitter Families — that is what we mean. We must encourage Fitter Families."

"Exactly!" enthused Harry. "That is exactly what I mean. Say, Nigel, I kind of like that name for a contest."

"Fitter Families for Future Firesides," Nigel replied, now smiling broadly at his clever alliteration.

"By jove, I think you've got it — the name of a whole new contest."

"I'd need some time to think that over, gentlemen," Miss Pinchot said, not wanting to appear close-minded. "Still, these ladies out here have grown quite fond of the contest as it is. The public health nurses are quite supportive as well. They might be slow to embrace a change."

"To be sure, but please give it some thought," Laughlin finished. "Perhaps we can have Nigel work with you and come up with some ideas on how such a contest might work. Fitter Families for Future Firesides — I really do like the idea."

(Egad, man! Better Babies… Fitter Families …! Better or fitter than what? Might as well be judging kumquats. How confident their stride, even as they were headed down the wrong road. They should have stuck to judging pies and cakes, or better still their own parenting. Lord help me … judging babies! As if they had anything to do with creation…

that is other than a few minutes of puffing and sweating. Aren't you even the slightest bit concerned?)

[**I AM.**]

(That's it? 'I am.' That's all you've got to say for yourself? Shouldn't you have done something?)

[**What did you expect me to do: throw lightning bolts, or tip over the judging table?**]

(Well, you could have done something ... anything. Perhaps you should allow me to intervene.)

[**Are you through? You do go on so. You seem so eager to slay the next dragon. Can we just drop it for now?**]

After leaving the exhibit, Laughlin suggested they stop by the beer tent again to refresh themselves before heading back to town. "Nigel, I've got to tell you that was a fine idea. The baby contests were a good start, and the local society has done a good job with their little pavilion, but we need to move them along. Gad, measuring intraocular distance and taking measurements of a child's cranium are so crude. Marking a baby down for squalling and evacuating, all while displaying him in his altogether, and that while on a tabletop in front of a crowd of crowing ladies. And in this God-awful heat. It is beyond the pale, don't you think? Babies cry and soil themselves. That's what they do, isn't it?" he said with complete disgust. "We need to get at rewarding a whole line of acceptable family traits. That is what will get us to our goal. I think this might be a good project for you to take on — that is, if you are willing."

"I would be willing," Nigel said without the least trace of reluctance. "Once we obtain a little more momentum enlisting the support of hospitals and training schools to adopt IQ testing, I can see how I might devote more time to this. Meanwhile, if you are agreeable, I will give it some thought, and share my ideas with you."

"That's a good fellow," Laughlin replied. "Now, let's finish off with a couple of beers, then head back to town. We can grab a bite to eat there, where prices are more reasonable. We'll head back to Winfield tomorrow morning. We can say our goodbyes before catching the train back home. I don't know if I told you, but I will be meeting with Professor Hayes again. I've got a few more things to go over with him before we leave Chicago for New York."

Harry and Nigel arrived back at Winfield just after noon the next day. They had a short visit with Dr. Case. The purpose was to recap their work of the last week and to discuss future plans, particularly as they related to the revised sterilization bill, which was at the time stalled in the Kansas legislature. Nigel excused himself from their meeting to go over to the children's day room to meet briefly with Julia. Julia was just finishing up a session with the parents of one of the inmates. He waited just outside her office until they departed, smiled pleasantly at them, then entered. Julia was putting a last comment on the Record of Family Traits form when she looked up to see him standing in front of her desk. Her nervousness at his sudden appearance was unexpected. Given the ease with which they had formed a bond on the trip out to Winfield, and their week's work together, he thought to himself that she should have been much more relaxed. Nigel took pains to put her at ease with a casual remark about his observations on the Kansas State Fair. She appeared to regain some semblance of composure, and came around from behind the desk. She stood in front of him, but turned shyly away to avoid looking directly into his eyes. Not certain what to do with her hands, she straightened up the files on her desk, then looked up at him and asked how the contest went. Nigel made an amusing comment about the process of cranial volume calculations he and Laughlin had witnessed, then offered vague suggestion of plans for the contest's improvement. The generalities in which he spoke suggested to her that he had something else on his mind. She made it easy for him to open up about whatever it was he was thinking by inquiring whether he thought she was ready to carry on in his absence. His face brightened, and she knew then that he was prepared to speak more, this time more directly. Although his assurance of her readiness was offered with kindness and unrestrained regard, it was his gestures that conveyed more of what he was thinking, or perhaps feeling. For even as he spoke, he drew closer to her, so that he had to look down into her eyes. He started to raise both hands as if to reach for hers, then pulled them back. He stepped a little away from her, paused, then said that he had the utmost faith in her. He went on to say that she should write him at the laboratory, and that if she had any concerns, any whatsoever, she should feel free to openly express them to him directly. Just before their hesitant parting, he assured her that they would no doubt work together again, perhaps in less than a year. Haltingly, he drew nearer once again, this time to shake her hand, and offered his final good wishes. Touched by his assuring words, she leaned in and gave him a kiss on the cheek. Blushingly, they parted.

CHAPTER XI

JULIA FINISHED HER ASSIGNMENT at the Kansas State Home for Feebleminded in the summer of 1917. During her time there, she and Nigel corresponded regularly. She usually included a personal note with her regular performance reports. In his last reply, he noted that she had completed an astounding number of Family Trait forms, and administered well over 200 tests of intelligence. Owing to her accomplishments, he arranged for her return to Cold Spring Harbor so that she might help with the administration of the new Army Alpha and Army Beta tests. Those tests of general cognitive ability were designed to rapidly assess a massive number of new recruits.

By age twenty-three, Julia had been exposed to more social inadequacy than might be expected for even the most experienced eugenics field-worker. Over the year following her return, she and Nigel, along with a large cohort of trained military and civilian clerical workers, helped administer the Army intelligence tests to nearly two million young men. Their work ultimately produced a trove of data which would further the aims of the eugenics movement. Coupled with Goddard's assessment at Ellis Island of immigrants' intellectual capacity, eugenicists accumulated enough data in those years to more concretely inform public policy.

Harry Laughlin and Charles Davenport naturally thought of themselves as guardians of the most salient populations statistics on intelligence at the time. Their data stores offered irrefutable proof, on a massive scale, of what they had long privately considered a settled question about the incapacity of so many to function in modern society. Even as early as 1913, owing to his work with immigrants, Goddard had proved to himself that many immigrants, particularly those from southern and eastern Europe, were poorly suited to the demands of life in America. No longer was social inadequacy a supposition; it was for them a proven fact. As to determining the intellectual fitness of the great mass of Americans, it was no longer a matter of subjective observation. People could be scientifically sorted using only a simple paper and pencil test, one that could be administered in less than an hour to groups as large as 500 persons.

It was not surprising that in 1920, bolstered by the accumulation of these scientific data, Harry Laughlin appeared before the House Committee on Immigration and Naturalization to testify on the biological aspects of immigration. He once said in response to a question from Representative John C. Box of Texas, "It is doubtful whether there is a single country in the world that does not have many families so splendidly endowed by nature that they would not make excellent and desirable additions to our citizenry. But because our foundation stock is largely from northwestern Europe and our national life was largely determined after the Northwestern Europe pattern, we find the assimilation of immigrants from this section of Europe to be a much simpler task than the Americanization of Latin or other stocks less closely related to us in nationality. We like to think also that the percentage of hereditary excellence is higher in our parental countries of Europe than in other nations. Perhaps it is; but by setting up a eugenical standard for admission demanding a high natural excellence of all immigrants regardless of nationality and past opportunities, we can enhance and improve the national stamina and ability of future Americans." He went on to argue that inferior individual family stocks, and not just inferior nationalities tended to deteriorate desirable national characteristics, which was, in his view, a serious national menace. Thus in 1921, the new Emergency Immigration Act reduced quotas from southern and eastern Europe. In justifying this legislation, Laughlin claimed, "Every family, people group, and nation has a right to set up its own biological standard, and to act in a positive or negative way to enforce the standard."

(Bloody Hell!)

[Must you use such language?]

(Every time I hear that line of Laughlin's I get collywobbles. Maybe we should have made him an offer.)

[An offer?]

(That he couldn't refuse.)

[Need I remind you, that's not how we operated then, and it shall not be in the future. For the last time, please be silent and allow me to continue.]

Although Laughlin first published a suggested sterilization law in 1914, it wasn't until his book, *Eugenical Sterilization in the United States*, came out in 1922 that the practice began to grow dramatically. "Should it be at last more widely practiced," he preached at the time, "the scourge of the socially inadequate, characterized generally as degeneracy, delinquency, and defectiveness, might finally be eliminated." More specifically, he defined social inadequacy to include the feeble-minded; insane, (including the psychopathic); criminalistic (including the delinquent and wayward); epileptic; inebriate (including drug-habitués); diseased (including the tuberculous, the syphilitic, the leprous, and others with chronic, infectious, and legally segregable diseases); blind (including those with impaired vision); deaf (including those with impaired hearing); deformed (including the crippled); and dependent (including orphans, ne'er-do-wells, the homeless, tramps, and paupers).

Owing to his continued testimony before Congress, immigration laws were modified again in 1924. Quotas were reduced further using as a base the immigrant share of the population in 1890, when immigrants represented a much smaller percentage of the population.

(A man on a mission!)

[A man of science.]

(Lest we further annoy the reader with our commentary, what of our hero Nigel and his family? Pray, do tell what they have been doing all of this time?)

[Do forgive me. The context is relevant, as the reader will in due course be given to understand. In any event, Nigel was no mere cog in the works, as you shall see. Do now hush up and allow me to continue.]

By 1926, the work of the ERO had become more than just a social science laboratory to tally and archive Records of Family Traits and Individual Analysis Cards. It was a veritable machine for the production and dissemination of facts and scientifically informed recommendations through publications designed to influence public opinion and decision makers. It was a public relations dynamo that lent expertise and guidance that would ultimately shape global public opinion on the necessity of, and the manner in which, all human stock might be improved. Nigel had labored long in the field, collecting and cataloging the wide variation of human affliction, sorting, measuring, and analyzing anomalies of nature. By 1926, Nigel's work transitioned to curating the trove of anatomical, temperamental, and intellectual properties that comprised the totality of the human family. In sum, it was his job to keep and share the records in such a way as to influence those who had the power to get on with the imperative of remediating the human condition. This was no small order for our Nigel, a man with nobility of purpose, formed so long ago in the crucible of higher education. Yet cracks were beginning to form in his commitment to the movement.

Fortunately for him, he could draw satisfaction from his reliance on a stable and supportive helpmate: Gemma, who had endured his many absences over these years, all the while nurturing two small children into sturdy and satisfactory youngsters. She had kept the home fires burning, shouldered her wifely duties without complaint, and stroked the too often fragile ego of her husband. It was a testament to her own forbearance, no less than her sagacity, that she kept her personal sentiments about her husband's work to herself, and managed to raise two pliant, uniquely endowed, and loving children.

Gracie blossomed into a talented, formidably intelligent, and brightly optimistic young lady of eighteen. Well advanced in her studies, she looked forward to choosing from a number of offers to attend elite schools of higher learning. Of course, her father sometimes overbearingly encouraged consideration of one of the finer Eastern women's colleges, especially those with which he was most familiar. Gemma, on the other hand, considered the choice to be one that could best be made by her daughter in the absence of any parental guidance. She had admonished Nigel then that Gracie's interest in the sciences, and medicine in particular, might suggest an altogether different choice than one he may have had in mind.

(And what of the boy? So much time has elapsed since the last time he was mentioned in our story. You ought to tell how he had progressed.)

[**Ah, yes, the boy.**]

Eugene grew into an eager and active child, one who delighted so many through those early years. It was for a time as if he had never discovered his own limitations. Though later, as a young boy, any such limitations were not of any unusual sort insofar as getting on with daily life, they were instead limitations in his ability to handle tasks that required consideration of future consequences. That is to say, he lacked sufficient executive function to anticipate more abstract cause and effect relationships. Thus, he could find himself in jams from which he was unable to extricate himself, such as the time he climbed a ladder onto the roof, and once there, hoping not to be caught, he pushed the ladder into the shrubbery below. While others could see his disabilities, he was altogether oblivious. With supervision, he simply went about his business, doing what he could do as if he lacked ego, and any sort of competitiveness. It wasn't until his own incapacities required externally imposed constraints that he acted differently.

As to the discovery of such limits on his capacity to act freely, he too often resorted to manifesting his displeasure in the form of flight. And to flight, especially rapid flight, he so often did turn. By age six, he had begun to run, even with short appendages that somehow only just managed to reach the ground. Whatever the trigger or whatever the impulse might have been, our young Eugene imperiled his own life at times by running hither and thither. Whether to town, the boarding house of Ardis and Cork, or to the laboratory, Eugene might too easily disappear without notice. I suppose it relieved tension, but it also had become for him somewhat of a game. Invariably, Gemma, Gracie, or a host of friends and acquaintances were deputized and sought him out in sometimes the most unexpected of places. One time, at the age of ten, he almost managed to talk himself onto the local train to Grand Central Station in New York. It was no small source of terror to his family, but thanks to the engagement of the townspeople, who by then had grown quite accustomed to the young lad's antics, he was found and safely returned to his family.

His running began, it can be supposed, as evidence of his desire for freedom. At first he ran because he discovered that he could, and he ran with sheer joy, bursting all the while in peals of laughter. Though vexing to family, it had become a matter of great amusement to the townspeople who had seen it.

But as he learned more about himself in relation to the world, a world that was sometimes cold and inhospitable at best, even at times hostile, Eugene's running took a new form. As he grew older, he had begun to sense that the world would rather squelch his freedom, and perhaps, though he could not put a name to it, wanted to sequester him. He did not hear, or did not comprehend, the backhanded and subtle comments that were made in his presence. But he could discern an aura, a cloud of prejudice that shadowed him. Too often, those who did not know Eugene walked to the other side of the street, or shielded their children on his approach. Such gestures did not escape his notice, even when his own family was present to distract him.

Eugene, like any other sentient being, when confronted with threat, had the option of fight or flight. Without the slightest inclination to aggression, Eugene of course opted for flight. And so it was that his running had changed into retreat, though never in panic. He had long ago learned that his own physical attributes, limited though they may have been, were sufficient to remove him from any apparent threat. As he grew and learned still more about himself in relation to the world, and with Gracie's constant and loving tutelage, he began to develop the verbal skills and temperament that could deflect threats that had once caused him to retreat. He learned that a smile, laughter, and simple, pleasant words could often disarm those who would inflict harm or try to avoid contact. Oh, and a generous and unexpected hug always seemed to do the trick. No more was the world changing Eugene; he learned that through dint of his personality he could change the world around him. And so he did.

Curiously, at about the time these changes were taking place, he began to walk on the balls of his feet, almost on tiptoes. There was a new bounce in his step, as if he were walking on clouds. He carried himself with full extension, casting his eyes about to continually see and sense the world, an ever-present smile on his lips. Surely, it was amusing to observe, being unexpected, unconventional, and always disarming. It was, to those who knew him, yet one more endearing quality that in time made him not only a familiar and accepted sight, but one of the town's very own. His presence lifted the pall of daily toil, let in sunlight, and generally made the simple folks around him feel good.

Lest one think that Eugene was some manner of a lesser god for all of it, he could be obstinate, and at times so wrong-headed that he exasperated even those, or perhaps especially those, who loved him most. Like most people, however, he found such willfulness simply a way to establish a boundary between himself and others. He was his own person, and no one was going to deny him that.

By the time Eugene was fourteen, he had, through Gracie's diligent efforts, completed the equivalent of grade two. He could write most letters correctly if sloppily, possessed rudimentary arithmetic skills, and could navigate most familiar signs. It was true that much repetition was required to achieve this level of academic advancement, but by this time it had become apparent that additional academic work would not be required for him to achieve a sufficient degree of satisfaction and independence in life. For it was his personality that in the end would make him into the kind of person who could be fulfilled — that and a sometimes stunning level of unexpected wisdom that even the most thoughtful person had to acknowledge. As Eugene once observed to his father, "Most people are afraid of others who are different. They try to hide it, sometimes by making fun, sometimes with bad words, and sometimes with trying to be especially nice. But the people who are different can see it, almost always."

When he was fifteen, Eugene discovered an exceptional capacity to work. He was relentless in finding a job. Work, in and of itself, would provide self-satisfaction. It would be better still if in performing well he pleased his employer, and to a large degree he did just that. He found that stocking shelves and generally cleaning up around the local grocery story suited his talents. He could greet and talk with customers, and rely on a repetitive daily routine to complete his assigned duties without much supervision. So accomplished was he in learning the preferences of the store's regular shoppers, that he was often able to select and bag their desired items before they came into the store at their regular time each week. Even the proprietor had to admit that no other store clerk he ever employed showed such a level of commitment and positive engagement with customers.

As if this was not remarkable enough, Eugene often demonstrated an uncanny ability to offer constructive advice to his customers.

Once, he was overheard by the store manager telling a regular who had too often feuded with his neighbor that he should not have hate in his heart for his neighbor, but that he should buy him an apple pie and take it to him that very day. On the surface, this seemed to be simply naïve, if not stupid advice. So simple it was, and delivered with such an unassuming good nature, that the crusty old customer was persuaded by Eugene's positive persistence to give it a try. Perhaps it was only to be rid of Eugene following him around the store, but the customer made a special point to tell the proprietor the next time he came in that things seemed to be going better with his neighbor, and that he might just have to order another pie.

It would be misleading to say that, aside from occasional stubbornness, all was sweetness with Eugene. He could be sullen and temperamental at times, but these moods were seldom. And when he was, he invariably got over it quickly. Mostly he was just lovable Eugene, and nearly everyone had taken a liking to the young boy – everyone, that is, but Harry Laughlin. Although Laughlin only frequented the market on occasion, and then only with his wife, he made it a point to avoid any encounter with Eugene. He had reasoned with his wife that it would not be fitting to engage with the child of an employee of the lab, but that was not his real reason. Rather, it was that he still harbored strong feelings that imbeciles such as Eugene, and perhaps especially because Eugene was the son of an employee of the ERO, should have by age fifteen most certainly been institutionalized. It just set a bad example. In any event, an imbecile ought not to have been seen in public, and most especially in a place of business.

In those intervening years, to Davenport and Laughlin, Nigel seemingly developed into a most reliable acolyte. Perhaps it would be more accurate to say that he appeared to them to be wholly dedicated to the movement. Although he would hew to their priorities and predilections, he had become so committed to advances in the science of general cognitive ability that he attained some measure of independent thinking in this area. It was becoming the new quest in his life. For Nigel had come to realize that all manner of degeneracy was most often in some way linked to substandard intellect. To him, it should have been obvious to his peers, but they were so focused on familial and heritable patterns of general social inadequacy that they could not see past the imperative of limiting reproduction. Their differences had by this time become a source of internal tension for Nigel, one he of necessity had to mask.

Having by then become familiar with the Stanford-Binet Test, and having fully immersed himself in the wealth of findings from the Army Alpha results, he became so fixated on the topic that he decided to test his own children. That he should have done this was not unexpected, given the obvious discordance between his and Gemma's own native abilities, versus those he believed were inherited by Eugene. Especially perplexing was the dramatic variation apparent not only in the physical characteristics, but also the intellectual limitations of his son.

His finding of an intelligence quotient of 145 points in Gracie and 42 in Eugene did not merely confirm the obvious, it served to put an important scale to the difference. How he might respond to the results was a matter of some deep consternation to Nigel. Many nights were lost to brooding on the problem. But then as we know, Nigel had always been given to mental fixations. In the end, he concluded that it might some day be possible to not only arrest intellectual decline, but that intellectual differences might be altogether avoided through some as yet to be discovered form of biological intervention. Such intervention, he thought, should not have to rely on any form of regulated or managed breeding. The latter approach, he had reasoned, was too slow and too imprecise to alter the path of civilization, let alone one's own family. This, he realized, was heretical thinking for a contemporary eugenicist, so he kept these ideas to himself. Still, he knew that given the advantages that intelligence confers, it would be desirable to in some way be able to somehow alter germ plasm itself so that intelligence, at least above some base level, might be achieved for everyone. Of course, were it possible, one would thereby also achieve a diminution in the rate of other forms of social inadequacy. So lost in thought about such things was he one evening that, when Gemma jostled him out of his deep rumination, he recoiled with a jolt.

"Nigel, dear," she said to him trying not to arouse alarm, "I have something important to tell you."

"Hmm? Oh, what is it, Gemma dear? I was just thinking about something. Sorry," he replied, sounding as if he had arisen from a deep slumber, which might be rightly supposed from his then complete detachment.

"I have found a lump. I'm not sure it's anything, but I confess to being somewhat troubled," she admitted.

"It's probably nothing," he said to reassure her. "Still, I suppose you might have Dr. Abraham take a look."

"Yes, I suppose I should," she said. "I don't mean to alarm you. It's probably nothing, but it would be irresponsible to neglect it."

"Quite so," he said. "If you like, I can ring him up in the morning."

"That's all right. I'll just go in. I've got some marketing to do tomorrow."

"Do let me know the outcome. I'll be at the lab tomorrow. You can get a note to Miss Chalmers if I'm not available."

◆

Gemma was diagnosed with breast cancer. After two radical surgeries and a prolonged illness, she succumbed to the disease later that year. For the latter part of her illness, Nigel was temporarily relieved from his duties at the lab, and stayed at home to nurse Gemma and care for the children. Gracie finished school, Eugene continued his work at the grocery store, and they all leaned on each other for support more heavily than normal. Ardis and Maria came by each afternoon in the later stages of Gemma's illness to give Nigel some relief, and Cork aided by doing various repair and maintenance jobs to the cottage. When Gemma passed away during the summer of 1926, she was buried in the cemetery that overlooked Cold Spring Harbor.

Gracie deferred her plan to start college in the fall, having earlier selected Radcliffe because of its association with Harvard. She insisted that caring for her father and brother were more important, and though initially disappointed, made peace with her decision, despite her father's insistence that she continue her schooling.

Eugene was emotionally lost for a time, but seemed to regain his footing because he was able to rely on the support of the store's proprietor, and the emotional support from Gracie and family friends.

Nigel returned to work right after the funeral, and was given plenty of space in which to grieve. He found the distraction of work a great comfort, and used the demands of his job to suppress the grief that might have otherwise crippled him. For the most part, Laughlin was supportive, at least in an economic sense, having continued to provide a salary during the period of Nigel's extended absence. After allowing what he considered a sufficient time to recover, Laughlin did, however, become a bit more obvious in his suggestion that circumstances now more clearly warranted Eugene's institutionalization. Needless to say, these more direct observations bordered on personal intrusion at a time when emotions were raw. At the same time, Nigel began to have growing contempt for the prospect of imposing on society some impractical notion of eugenical marriage and procreation, it being in direct contradiction of primal forces and baser human tendencies, and for that matter his own.

That Nigel had managed for so long to repress these emotions ultimately made the situation explosive. Absent Gemma's steadying words, Nigel no longer had the emotional ballast to ride out the storm. In time, harsh words were exchanged. Pride was grievously wounded in both men, and what had been an uneasy truce for more than a decade erupted into emotionally destructive warfare, the terminus of which could only have been, for Nigel, flight. And in a fevered search, Nigel hastened to find a letter he had retained for no particular reason other than he had an unexplainable attraction to his conception of the state in which their correspondent resided. It had been suggested by the author of the letter that Nigel should consider coming to work with him. At the time, such a move would have been impractical. More than anything, though, it had been as much the mere sound of the name, so exotic, so distant. Having never been there, of course it seemed irrational, yet there was something in the name, something ... Iowa. So to Iowa he did indeed retreat. He was there to be retained in an administrative capacity at the Iowa Institution for Feebleminded Children.

(Hold on! You're moving kind of fast – skipping over things. The reader by now has got to be asking: Is this going somewhere?)

[**They need to stick with me just a little bit more. As you know, I've a lot of ground to cover here, and not much time to get to the point.**]

(The point. Ah, yes ... the point. What exactly, they must be thinking, is the point?)

[**Again, they need to stay with me. We're near the end ... of the beginning, as one shall see.**]

(The end? Near the end? Will we ever be near the end? I think we might have done a bit more editing of this tale.)

[**Tut, tut. As you well know, I would consider any alteration of this story to be a dastardly act. In any event, I'm coming to a** point. As I'm certain you well know, when it comes, our readers will think it has come too quickly.]

CHAPTER XII

Nigel paused in the circle drive to look up at the grand front entrance while he waited for Dr. Mogridge to park the auto. The looming edifice reminded him of a bird, the Jaeger that he had once observed on a visit to the English coast. With neck and bill extended toward the heavens, it balanced on one foot atop one of the many pilings rotting in the sea. Barred wings extended fully from its body as if it was ready to take flight. The bell tower sounded the time, jarring him from that recollection as his host commented, "Impressive, isn't it? The architecture, I mean."

"Oh yes, quite," Nigel replied, shaken from his reminiscence. "I was not expecting anything quite so grand. Uh, perhaps I imagined something a bit less imposing. Yet in its own way it's quite beautiful. A Kirkbride, no?"

"It is. You are familiar with the concept, then. Its design offers a sense of decorum, dignity, and even permanence, don't you think? The architecture is thought to have a curative effect. As you can see, the building extends up four floors, and sweeps symmetrically away to both sides, creating an atmosphere of order and privacy for the inmates. The dormer windows are mostly just for show. The same for the bell tower, though the bells, as you can hear, are quite functional, and in a way govern many of the activities here on our campus. The bars on the windows are merely for the inmates' safety. Shall we go in — perhaps take a quick tour of the main building?" inquired Mogridge.

"Yes, that would be lovely. I am anxious to see inside, if it's anything quite as striking as the exterior," answered Nigel. On arriving in Iowa by train, he had already thought that the state's rolling, pastoral character created an atmosphere much more conducive for sheltering the enfeebled than had Kansas' rather harsh, featureless landscape. As to his prospects for working there, he was beginning to consider it quite encouraging.

Dr. Mogridge had been superintendent at the Iowa Institution for Feeble Minded Children in Glenwood, Iowa since 1903, and was its third director. He had been formally introduced to Wellbourne the previous year at a eugenics conference in New York, though he was already familiar with Nigel's contributions to *Eugenical News*. They had been amiable correspondents over the months since their first meeting. Mogridge perceived in Nigel the energy, dedication, and like-mindedness that might one day give them a good foundation for a possible professional relationship. It was not surprising, then, that shortly after learning of Gemma's passing, Mogridge suggested in a letter to Nigel that he might consider employment at the hospital should he wish to make a fresh start. Their meeting at the institution was to be an opportunity to confirm Nigel's interest in working there as well as a chance to discuss ways in which Nigel might apply his own unique skills to the growing institution. Nigel's visit was important enough to Mogridge that he had arranged to pick up Nigel personally at the train station and devote a full day to discussing his possible role.

Nigel considered the possibility of a move only in the last month, after first discussing it with Gracie. Inasmuch as Gemma had been ill for an extended period, and had only been departed for three months, Gracie had still not made a final decision on when to start college, and whether to continue with her plan to attend Radcliffe. The prospect of a move to the Midwest was timely, inasmuch as Gracie had once considered one of the Midwest's land grant colleges rather than an Eastern school. Before considering medicine, she had thought of pursuing a course in general biology first, and then taking advanced coursework in animal science. The prospect of her move caused her to alter her plan to go to Radcliffe. She decided to apply to several Midwest colleges instead. On the basis of her strong academic record, and with sterling recommendations from her secondary school teachers, she was admitted for the winter term to the University of Nebraska, where she was given a generous scholarship. She would take the intervening months to help her father and Eugene get settled.

After a tour of the farm complex, which comprised nearly 160 acres and numerous outbuildings, Wellbourne and Mogridge returned to the main offices for more detailed discussions. While walking up toward the main campus, Nigel made an observation about the farm manager that all but confirmed for Mogridge that Nigel possessed the keenest insights. To the director, this would seem to make him an ideal candidate to head up the institution's psychology laboratory.

"This Bob Horn chap, is he successful in what he does? That is to say, does he accomplish all that might be required to balance the needs of the farm with support for the inmates in his care?"

"That is," Mogridge replied, "a most perceptive question. I suppose it arises from your observation, however diplomatically put, that he is not the most patient of men ... at least as it concerns his interaction with the boys under his supervision. Still, he runs an excellent farm. As you no doubt may be aware, institutions like ours depend in large measure on farm operations for self-sufficiency. He has done quite well in that regard. So well, in fact, that he has been especially called out by our board and the Department of Agriculture for his highly productive herd of Holsteins. Of that, we all are quite proud."

"That is commendable," responded Wellbourne. "But is it, in your mind, sufficient? That is to say, might it not be possible to run an acceptable farm while also providing support and development to those in his charge?"

"You've put your finger on a key issue. And that is why, especially, you are here. As you know, I've followed your work for several years now. It was our meeting last year that confirmed for me that with your expertise in testing, and your personal circumstances ... er ... with your first-hand knowledge of ... of those with limitations, that you might help us sort out just how we might better assign our inmates to school, and correspondingly to work. With respect to each group, we would like to learn whether and how we might more effectively influence their development."

"I think I take your meaning, but ..." Nigel hesitated, and then started to complete his thought when he was interrupted.

"Let me try to be more succinct. The question in my mind, and in the mind of our board, is whether intelligence can be improved in some way, so that some day we might not have to, as a society, bear the burden of long-term care. I myself have grave doubts. Nonetheless, it is imperative that we answer the question through careful research. To date, we have done little more than assume responsibility as a surrogate family or caretaker for such children, and that comes with all the costs associated with such care."

Their conversation had taken such a turn that both paused in their walk up to the main building. They stood silently on the highest point of land, and looked out over the Missouri River Valley and the verdant farmland fading to the horizon. It cannot be said with certainty when Nigel, long in the counsel of thought leaders like Pearson, Laughlin, and Davenport, had already realized that procreation could neither be compelled nor constrained from outside the bounds of connubial bliss, and thus the eugenics movement was, as originally conceived, on shaky ground. Rather, he was then beginning to contemplate ways in which care for the disabled might be improved, particularly with respect to their intellectual development.

"Yes, it is an intriguing question," Nigel said. "It has long been on my mind. If I may, please allow me to share something confidential with you. As you know, I have worked for a particularly driven man. Were it not for him, the eugenics movement would not be as far along. That said, I have grown increasingly concerned over these last few years with a movement that at once tries to do what is right for humanity, but at the same time deprives individuals of their inherent rights. In particular, though it is not without cause in some instances, I have grave doubts about the rather broad application of sterilization or controlling birth rates as techniques to eradicate social ills. It seems to me ...."

Once again, Dr. Mogridge interrupted. "I am sympathetic to your concern, but I think you will agree that there are circumstances in which such methods are advisable. Of course, we must always consider the wishes of the family, and take a measured approach. As I am sure you are aware, states like ours have protections in place — in our case, we refer such cases to the State Board of Control for oversight."

Nigel reflected for a time on what it was that Dr. Mogridge was saying before speaking. He thought to himself that perhaps he had been too candid, and so opted for a more conciliatory response. "Yes, as I said, there are indeed cases that may warrant such measures. And it is decidedly good to have such means of oversight as you have stated. Iowa is to be commended for this."

Dr. Mogridge smiled faintly, and they resumed walking toward the administration building.

"I think that I must have been inarticulate in what I was attempting to say. Please do forgive me, and allow me to try again," Nigel said.

"Do go on, then," answered Mogridge.

"What I meant was that for the movement to make necessary progress, we must also examine ways in which the lot of those we institutionalize might be improved, with the ultimate aim that we might find a way for them to function without dependency on the state." Nigel felt that this time he had done a better job at explaining himself.

"Yes, and that is what we would wish you to discover through your research here, that is, if you are game. But let me just add that I have not seen thus far any evidence that sustained improvement in intellectual functioning might be achieved through any outside intervention. I say this not to discourage you. Rather, it is my hope you will one day find me wrong. Ah, here we are. Shall we go up to my office and talk further?"

Over the ensuing hours, they discussed Nigel's work history, reviewed his skills, and examined their respective philosophies on eugenics. Both judged there to be a near complete meeting of the minds.

"Well, then, what do you think?" asked Mogridge. "Do you think you can be happy here?" Before giving Nigel a chance to answer, Mogridge slid a written offer across the table. "I think you might take a moment to read this before you answer." Dr. Mogridge smiled broadly.

"A house. The position comes with a house?" asked Wellbourne incredulously. Not only was he stunned by the unexpectedly generous salary, but the fact that it came with housing made the total offer substantially more than he was making at the ERO.

"Yes, of course. The space was freed up when our former director resigned. Pity in a way. He was a bright fellow … a Ph.D. in fact. I hasten to add, such an education is not required for the position. It might even have been a hindrance. I find most Ph.D. types ill suited for the work. In any event, he was clearly uncomfortable among our population, and that just wouldn't do."

"That is very generous … with the house and all, I mean. It would be just me and my son Eugene … my daughter Gracie for a brief time," Nigel responded.

"Ah, yes, your son. You have spoken of him a number of times. A Mongoloid, I believe you have said. Perhaps, if it seems appropriate to you, once you are settled, you might consider enrolling him in our institution. He would, I expect, be happy and well adjusted among his kind. We could offer schooling or some sort of vocational training … whatever seems best. Again, there would be no pressure on you to make this happen. The choice would be yours, of course," Dr. Mogridge asserted.

"That is kind of you to offer. Eugene has done very well at home, and he has held a job before. That in large part owes to his mother and his sister. However, with Gracie starting college, and Gemma …." Nigel started to choke up, then began again. "Yes, I think we might consider it, but I will have to see after we get settled."

"Does this mean that you will accept our offer?" Dr. Mogridge inquired.

"Yes. Yes, it does," Nigel said with a broad grin. "Yes, I think we will like it here very much. I must say, Iowa seems a bit of heaven, now that I have seen it in person."

"Oh, one more thing. Your first order of business is to hire an assistant. It would not be possible for you to carry on the kind of work we envision without some assistance," Mogridge added. "The Board has been quite supportive in funding our research effort. I trust you have someone who might fit the bill."

"Indeed ... em ... I think I do," Nigel replied, still feeling giddy about his new prospects.

"Shall we take a look at your work space? I trust you will let us know whatever you might need. Then, let's run over and take a look at the housing we have for you. You will be pleased to know that it comes furnished."

◆

Nigel, Gracie, and Eugene made the move to Glenwood, Iowa, in September 1926. The transition went smoothly, save for the expected delays in rail service. Harry Laughlin took the news with equanimity, having for some time reasoned that the need for Wellbourne's supervision of field-workers had passed. Although some data collection efforts would continue for a time, the primary work of the lab shifted to influencing public policy, work that was better suited for himself and Davenport. Moreover, the board had begun to challenge the growth and absolute size of operating expenses. Perhaps it was less the expense than the manner in which Laughlin had staked out what many considered to be rather extreme positions on immigration and sterilization, particularly given emerging resistance to his aims. Cost containment was just a convenient and non-threatening way for those who funded the organization to rein him in a bit. Given the board's prominence in the broader affairs of the nation, it was best that the ERO begin to lower its profile. To that end, responsibility for the publication of *Eugenical News*, which had likewise begun to reflect

Laughlin's more strident views, was shifted to the parent, the Carnegie Institute in Washington.

When their move was completed, Gracie used the time prior to starting college to explore Glenwood with Eugene. This was to assure the ease of his transition, anticipating that Eugene might otherwise have a hard time adapting to his new surroundings. Although it was highly unexpected for a Mongoloid to be seen out and about, especially in the town that was home to the state's institution for feeble-minded children, Eugene adapted quite nicely. His cheerful countenance and outgoing personality soon gave him some degree of celebrity, at least among the shopkeepers and townsfolk they regularly encountered on their daily forays into town. Given that their home was on the campus of the institution, which was on the edge of town, their daily walk assured their physical as well as mental health, and the two siblings used the togetherness to great advantage. Eugene began to transfer his affection for and dependence on his mother to his older sister, for whom he already held great affection.

As far as their housing situation, the three were quite comfortable. The home was in substantially better condition and more up to date than the cottage that they occupied in Cold Spring Harbor. It had all of the latest conveniences: electricity, telephone service, and an oil-burning furnace. Save for the adjacency to its neighbor, the adjoining unit in their duplex, it was private, so they could enjoy family time without the judging eyes of those around them. Needless to say, it was out of the ordinary for a family with a feebleminded teenager to be living among the other staff of an institution whose very existence isolated those who by accident of nature or neglect possessed qualities that made them better suited for separation from society. Ever protective, Gracie was especially sensitive to any innuendo or non-verbal expressions of disapproval that might come from the institutional community. Nigel, who so easily became absorbed in his work, had no sense of the way in which his non-conforming family arrangement might arouse local sentiment.

Nigel immediately began to formulate a plan for his lab, which forthwith included the retention of able assistance. That would require an intermediate step, which was not to be without several complicating factors. He immediately renewed correspondence with Julia Benson. That alone was a source of minor tension with Gracie. Not to be fooled by the imperatives of her father's work, Gracie was more than sufficiently mature to see that the platonic relationship her father had developed with a younger co-worker ought to have demanded a degree of discretion that could not in these pending circumstances be assured. Given that they were new to the community, and a somewhat staunchly conservative Midwestern community at that, discretion was of considerable import to the goodwill on which this family, and to be sure all families, must depend. For unlike the great majority of townsfolk, and especially fellow employees, Nigel could not possibly have been seen as anything other than a foreigner. His accent, a signal to some of self-assessed superiority, did not fool the cynical and untrusting eye of these small-town middle-Americans. Nigel was in every sense a newcomer, an outsider, and one not so quickly to be trusted, certainly for anything less than a generation.

◆

Nigel met Julia Benson at the small railroad shack on the rail spur that cut through the edge of the institution. She got off the train on a crisp afternoon in late October. The leaves had begun to turn, and the sounds of geese were overhead, signaling the approach of another protracted Midwestern winter. For now, the brilliance of autumn shone with great promise. Nigel loaded her trunk and small suitcase into his newly purchased automobile, and drove the short way to his home. He assisted her out of the front seat, and the two of them carried her trunk and small bag into the living room.

"This is lovely," she said, effervescent as always. "And the weather … the leaves are splendid. What a difference from dank Long Island! The land is so beautiful, so vibrant with color. You must also have thought this a great change from Cold Spring Harbor," she said.

"It is, yes, and we have become quite used to it," he replied. "Please excuse me while I carry your things upstairs to your room. You and Gracie will be rooming together until she goes off to college in a few weeks. I won't be but a minute. Then we can sit and chat awhile over tea. Gracie and Eugene will be back shortly. They've just run to town for some groceries."

Julia took off her wrap and hat, and moved to the far end of the room to observe her new surroundings. Though the space was small, someone, no doubt Gracie, had done a fine job of adding a woman's touch. On the walls were hung romantic prints of familiar European landscapes. Dried flowers were displayed along with a variety of ceramic figurines on the tables near the couch. A comfortable upholstered chair was in the opposite corner. The lamps were shrouded in printed cloth to subdue what otherwise might have been jarring light. The kitchen area looked homey and sufficiently well outfitted for the preparation of wholesome meals. In a cramped vestibule off the kitchen was a small dining table with four chairs. A centerpiece of fresh-cut flowers adorned the table.

As Nigel made his last trip upstairs to deposit Julia's small bag, Gracie and Eugene burst through the doorway with peals of laughter at what had been some shared silliness. Looking up, Gracie glanced at Julia. Her expression changed at once. Though not an unfriendly face,

Gracie's expression conveyed both surprise and a measure of resistance. Julia was an expected but not altogether welcome presence. Eugene, in his usual way, rushed to the still standing Julia, smothering her in an enthusiastic embrace.

"What time did you arrive?" asked Gracie, now trying to sound hospitable.

"Oh, just moments ago," she replied. "Your father is just taking my things upstairs. It is so nice to see you both again," she added. Julia knew that her arrival would be uncomfortable for Gracie. Both understood that with Gracie starting school, it was necessary to have someone home with Eugene until he was adjusted. At the time, Gracie did not know of her father's plan to hire Julia as his assistant. Good reason or no, both recognized that her presence in the Wellbournes' new home was more than just a small intrusion. It breeched a family bond at a time not long separated from the death of Gemma. Although this new living arrangement had a rationale, it touched a number of raw nerves, and could only at best be accepted, not welcomed. Consequently, both parties took particular pains in the days remaining before Gracie left for Lincoln to avoid any possible conflict. With forced smiles, the level of self-conscious propriety Gracie and Julia practiced was stultifying. Thus it was that there were no disagreements, no disruptions, certainly no turmoil in the interlude before Gracie's departure. Even if there had been, Nigel would have been too preoccupied with his own concerns to notice.

Gracie said her goodbyes at the train station in early December, and departed for Lincoln in order to get registered for the new term. It was to be a parting of some permanence. Although she would from time to time return home, her separation was more than just physical.

Nigel and Julia understood that after a period of her settling in, and in particular establishing a more trusted bond with Eugene, she and Nigel would have to delicately broach the subject with Eugene of his possible move to the boy's wing of the institution. It was, to their surprise, not so difficult as expected. They carefully staged a sequence of visits, ostensibly to expose Eugene to the possibility of securing work on the institution's farm. This, they hoped, would make the transition not only easy, but eagerly anticipated. During his visits, he met some of the boys, played with them during recreation time, and took meals with them. He even spent the night on more than one occasion. Ultimately, it was his visit to the farm, and in particular the time he got to spend with the livestock, that would seal the deal.

Although Gracie had known nothing of these plans, she was reassured through a series of letters that Eugene was adjusting well to Julia's presence. Gracie grudgingly began to accept the arrangement as the only practical alternative. Nonetheless, resentment lingered over the notion that a woman scarcely much older than she would assume the role long held by her mother. When later introduced to the idea that Eugene might find employment on the farm, she at least grew positive about his prospects. It was just after the middle of January that Nigel told her what Eugene had said. "I want to move in with the boys," was the way he had put it. Gracie accepted that it was for the best, given that Eugene at least would get to develop more independence. Building friendships would be a bonus. She'd always been concerned that one day Eugene would have to be on his own, and could see that it was imperative for him to find a way to separate himself from dependence on his family. That he would be living so close, and that her father could see Eugene as often as he chose, made the decision all the more accept-able. Moreover, she assumed that this would mean Julia's services in the home would no longer be required.

Because Gracie was not one to hold back when it came to confronting her father with what might have been impertinent questions, she would ask, "Just what are your plans with respect to Julia? When will she move out, and was she planning to stay in the area? Or would she be returning to the East? " Her letters home became nettlesome.

Nigel's replies did not settle the matter at first. Ambiguity and dissembling were for him the only viable response. In reply to Gracie's interrogatories, he assured her that Julia would be staying with him only for the time necessary to find an alternate arrangement. That she would become his lab assistant only for the duration of the research project was an unwelcome revelation to Gracie. Years of working alongside a demanding boss at the ERO had taught him the benefit of using oblique language and withholding unpleasant news.

Eugene gleefully made the move into the boy's cottage, where he celebrated his first Christmas apart from the family. And owing to Gracie's obligation to remain in Lincoln over the holidays to carry out her responsibilities as a part time veterinary assistant, Julia and Nigel celebrated Christmas at home alone.

CHAPTER XIII

Right after Christmas, Nigel and Julia met with Dr. Mogridge to outline their research plan. Although Julia would not officially begin her employment until the new year, her attendance at the meeting was Dr. Mogridge's first opportunity to meet her. He had by then heard much of her background from Nigel. Mobridge was even more impressed after meeting her. Nigel was not without guile, at least when it came to anticipating how he might gain advantage in encounters in which he expected to come away with something of value. He had cleverly calculated that her presence would smooth the reception of his plans, which included a rather substantial increase in the psychology lab's budget.

Nigel and Julia were to continue the intelligence testing already begun at the institution, which had been halted when his predecessor abruptly left. Their objective, beyond that of appropriately placing inmates in school, work, or day care, was to create a pool of candidates for research purposes. They would select an experimental group of twenty inmates of varying ages for a new kind of intensive brain development exercises. Another group of twenty, the control group, was to receive no special attention, but would continue with their normal activities. Both groups were to be matched to the extent possible based on IQ, age, gender, and family background. The target IQ range was 35-50, a category that likely comprised the largest proportion of inmates at the institution. To the extent possible, subjects for the experiment were to be within an age range of 5-12 and have similar family backgrounds. As they were able, Nigel and Julia would review files, and complete family trait forms for any subjects whose family members had as yet not been interviewed.

As to the manner in which the experimental group was to be stimulated, a number of carefully designed, non-educational games and exercises were to be used in multiple sessions per inmate each day. Of course, this required more labor than could be accomplished by Nigel and Julia acting alone, so student interns from the local public high school would be recruited and trained to administer these special brain development activities. Small stipends were to be paid to the interns. Unlike traditional primary school students, the experimental subjects would receive no lessons in reading, arithmetic, and the like. Rather, they would practice several short-term recall exercises. In another training session, the subjects would handle geometric objects, then practice folding two-dimensional drawings into those same three-dimensional shapes. Each day, the group would also listen to movements from recorded symphonies. Then, the interns would read short stories to the assembled group. In individual sessions, subjects would practice fine finger manipulation by sorting small objects into bins with speed and accuracy. Nigel had reasoned that such activities could stimulate brain growth. From his reading of work being done in France, he believed the repetition of basic sensory activity helped stimulate nerve cell growth deep within the brain. The resulting physical changes would thus enable higher cognitive functioning, which in turn would cause subjects to be more socially adaptable. He had for some time posited that it may have been the lack of early stimulation, attributable to being reared in cacogenic families that resulted in degraded brain development, and thus a lower IQ. Anecdotally, he had based his hypothesis on the demonstrable growth in Eugene's development attributable to Gracie's active work with him.

"I have to commend you and Miss Benson, of course," Dr. Mogridge started. "I am impressed with your speed in putting this plan together. No doubt your considerable experience, the both of you, has prepared you well to this point. Not without controversy, the idea of enhancing intelligence of the feebleminded gets at the very issue that our board has directed us to pursue in our research department. That said, the ultimate aim is to determine how we might figure out a way to reduce the burdensome expense to the taxpayer of housing these people." He paused for his statement to sink in. He went on, "Let me be more direct. Nigel, you will have to pardon me for repeating all this, but the purpose of any research is not just a matter of academic interest. I myself have concluded, as I am hopeful you may have as well, that relying solely on Laughlin and Davenport's methods, including sterilization and other such voluntary restraints on procreation, coupled with limiting immigration, will ultimately fall short in the effort to stem the tide of the mentally deficient. That is not to say that these things are wrong. Rather, they are simply insufficient. It is a matter of national imperative that we somehow reduce the already large and growing population of defectives. We simply do not have resources enough as a society to fulfill our responsibility."

"We understand that point of view, and endorse it wholeheartedly," Nigel responded. Even as he said this he had reservations about prospects for a successful research outcome.

"Let me just add," the doctor said, "as I have said to you previously, I have sincere doubts about the ability to effect anatomical changes such as you have described. I rather think that with respect to intelligence, the die has been cast at birth. I believe we are an amalgam of our forbears, a very muddle of bestowed characteristics. From our lineage comes all manner of mostly fixed abilities, and such limitations that it may be our misfortune to have received. It is only for you to prove me wrong in that regard."

By early spring, the research duo was well on its way to completing the necessary data collection and study participant identification. Unshackled from the imperatives of an overbearing boss, as had been the case at the ERO, their liberation gave them a renewed sense of mission. Meanwhile, Eugene adapted well to living in the state hospital. Owing to his exceedingly good nature, he had become well regarded both by staff and his mates on the third floor of the boys' ward. Known for his ebullience and occasional clownish antics, he had become a favorite of staff and inmates alike. Those who like Miss Mays, his ward matron, knew the boys as individuals rather than members of a category, learned to appreciate if not occasionally be put off by his antics.

One incident in particular demonstrated Eugene's manner of playfulness. To be sure, his sense of humor could on occasion rise to a level that one might not have expected from a boy of his cognitive limitations. In this case, his trickery manifested keen observation skills coupled with masterful acting. While Miss Mays was making her early morning rounds, urging the boys to dress with haste for breakfast, Eugene arose from his slumbers and, with grand gesture and a loud moan, crashed suddenly to the floor at the foot of his bed. Prostrate, he commenced violently rigid contractions, as if electrified. This created excitement and panic among the other boys, who circled the convulsive corpus to watch, first in horror, then in glee. As might be expected, most could find amusement in anything that interrupted the ordinariness of days. Hearing the clamor, Miss Mays rushed to his side. She thrust her hand under his head to prevent severe head injury, and spoke words of comfort to the apparently unconscious Eugene. In the course of his twitching and flouncing, while Miss Mays attempted to sooth away this sudden affliction, he unexpectedly looked up at her, wide-eyed. With the broadest smile, and a finger pointed directly at her nose, "Got you! I got you Mithus Maathe. I got you!" The boys jumped up and down, and in peals of laughter alternately clapped and shrieked with delight. Some called out Miss Mays with childish ridicule. Even they, the so-called dull and imbecilic, managed to find the humor in Eugene's clever trick. Ever humble Miss Mays, so often the object of ridicule in her own young life owing to her obesity, a trait often associated with both physical and mental sluggishness, was never one to take herself too seriously. Affably, she shared in the laughter. She had to admit, and in fact did so publicly, that Eugene had played a rollicking good trick on her. In a way, it was one of her finest moments, and was one more proof of the character that so endeared her to the boys on the ward. Ever after, she was invariably surrounded by a coterie of boys who, like a cloud of witnesses, testified to her goodness.

Miss Mays was not like most of the others who worked at the institution. Overweight, homely, and largely devoid of friendships, she had for the better part of her life carried a certain ignominy among the locals. The school years were especially hard on her, but she had developed resilience and was possessed of a degree of compassion that one did not customarily see. In her early twenties, she had taken in an afflicted cousin who might otherwise have died from sheer neglect. Although the dear child ultimately succumbed to her rather serious physical deformities, the experience motivated Miss Mays to do more for those who were cast off by society. That is why, no doubt, she seemed to possess the ability to see the worth and individual dignity of those in her charge. This compassion enabled her to function over the years as a mother to so many young boys. It was a blessing, too, for the Wellbournes — especially Gracie and Eugene. For in those early months in Glenwood, Iowa, before Eugene moved into the institution and Gracie had gone off to college, and while Nigel was finding his way at work, Miss Mays had been the touchstone to which both the Wellbourne children turned to find their footing in a totally new environment. It was good for them that she, along with another staff person, occupied the other half of the duplex they shared.

Eugene began working on the farm shortly after moving into the boys' cottage. Like most of the boys, he was first tried out in the garden, where his ability to pull weeds and hoe would be tested. Those who were unable to master weeding or hoeing were permanently assigned to brick making. That required only the packing by hand of wet clay into wood molds, which were then passed to an older, more capable boy to stack in the brick oven. On rainy days, all the boys except those who worked with the livestock were assigned to brick making. It was work that most anyone else would find boring and unfulfilling, but it was work that was needed for the institution's expansion. Garden work, on the other hand, didn't require much more in the way of capability, merely the ability to distinguish between plants and weeds. In the early summer, as the new plants began to emerge, only the hired help from town were allowed to pull out and discard the plants that were too stunted. It took a better mind to discern which plants wouldn't grow well. Fortunately for Eugene, he showed that he was at least capable of weeding and hoeing.

Not long after being assigned to work on the farm, Eugene let Mr. Horn know of his interest in working with livestock, particularly with the dairy herd. Although it had taken Nigel's intervention with Horn to give Eugene the chance to try his hand at feeding and later milking the cows, he demonstrated such skill with the animals that his ability could not be denied. Most of the boys worked at a pace that seldom came up to one-fourth the productivity of the hired help from town, though the latter were sometimes too rough and impatient with the animals to get the most out of them. That was especially true of Horn's son. In contrast, Eugene's gentle hands, though slow, nearly always produced more milk per cow. Thus, he was chosen to work with the dairy herd ahead of others. Horn was forced to concede that contented cows yield more milk, regardless of whose hand pulled the teat. In spite of the fact that Horn felt he had been pressured to give Eugene a favored assignment, he seemed to bury his own resentment for the greater good.

Eugene wasn't the only one affected by Horn's decision. Horn's son, a slow-witted adolescent, who too often found himself on the wrong side of the local constabulary, had recently been pulled from school and dragged by his father under protest to work at the farm. The son's presence sometimes demanded more of his father's attention than was required when dealing with the inmates. Assuming an air of privilege, this petulant miscreant, this tyrannical man-child, unreasonably claimed all manner of preference. So it was to cause in him profound resentment that a Mongoloid, one so obviously his inferior, would be given the opportunity to milk the cows. And thus it was, in the absence of his father's oft-observing eye, that he would bully, berate, and even occasionally rain blows on mild-mannered Eugene.

Late one afternoon, Nigel went to visit Eugene on the farm. He was planning to take Eugene shopping for some new work apparel. Horn was standing next to the garden speaking with a small tour group from the local business community. The inmates were busy weeding and hoeing. Nigel hung back, so as not to disrupt the tour. As Horn was answering a question, Eugene repeatedly asked Horn's son if he could work with the cows instead of weeding.

Hearing the disruption, Horn Sr. said, "Excuse me, Eugene. I'm talking over here." Turning back to the tour group, Horn Sr. continued, "Sir, what was your question again?"

"Your people," the businessman began, "do they get bored?" he asked.

"Some of your high-grade …," Horn Sr. began, before he was interrupted once again by Eugene, who had turned his attention from the junior to the senior Horn.

"Mithter Horn. Mithter Horn. I want to feed the cowths," Eugene insisted.

"Not now, Eugene. Can't you see I'm in the middle of something?"

Eugene stood his ground. "I want to work in the barn," he persisted.

"No, Eugene. You're going to weed today," Horn's son commanded.

"If you think I'm going to weed any more, you can kith … you can kith my, my, my ath!" Eugene stormed. "I'm bored with weeding," Eugene insisted defiantly.

The other boys stood up from weeding to see the developing brouhaha, ever ready to be spectators. "Come," Horn gestured toward the barn, trying to distract the tour group from an emerging insurrection. "Let's take a look at the dairy barn," he said, gesturing for the group to follow. There were a few titters from the back of the group, even as the group's leader did his best to keep a straight face. Horn's son, who was supposed to be making sure the boys hoed without damaging the plants, stood off to the side holding a shovel and glaring at a defiant Eugene.

Nigel, who had observed the whole scene, caught up with Horn Sr. who trailed the businessmen as they were walking back uphill to the administration building. "Look, Horn, I'm sorry about my son's behavior. I'll talk to him, and if it's okay with you, I'll take him into town right now to get him some work boots and gloves. That'll give him a chance to cool down."

"Yeah, that'd be best. The boy's got a smart mouth, and needs to learn some lessons. Best you handle it. But he better fix his attitude," he added. Horn turned away and hollered at his own son, "Get those boys back to work, and see that they don't damage any more of the plants!" Horn's son echoed this command back to the boys.

Nigel stood brooding for a time. He knew he had read Horn correctly. Horn was a man of limited patience, at least when it came to people. He was accustomed to having things his own way. Farmers were like that, Nigel thought. They can be awfully independent-minded. What can you expect from someone who works with domesticated animals? Animals don't have self-conception, let alone conscious will, and certainly not any kind of insight. All it took was a prod to get them to do what you wanted. So when it came to defiance, neither Horn nor his son was accustomed to being challenged. Nigel thought to himself that Horn certainly was not the ideal person to work with young boys — especially those whom Horn considered inferior. These boys required a special kind of handling. To Horn, maybe even the director, the efficiency of the farm was more important than sensitivity. That's why Horn still worked there.

At this time, the beginning of the summer of '27, to Nigel's great dismay, Julia unexpectedly announced she would have to return to her home in Massachusetts. A sickness in the family necessitated this sudden turn of events, she said. The absence of detail in her explanation and the suddenness of this announcement suggested to Nigel that some other issue must have been at play. Despite his best efforts, he could not get her to explain herself more fully. Julia was insistent that there was nothing else, but his persistence only added to the tension. So that she might leave on good terms, she went out of her way to show Nigel her gratitude during the final two weeks of her stay. Fortunately for him, they had completed enough of their preliminary work at the lab that he would be able carry on without her help. Julia had been effective in recruiting local students and church ladies whom he could train to administer the special brain development exercises. At the end of the year, he and perhaps someone else he could train would administer IQ tests to both the experimental and control groups to see if any significant changes had occurred. Even without those final IQ tests, Nigel could see that his research work was failing to produce the results he had hoped for. Increasingly disillusioned, he began to doubt the course of his life's work.

Living with Eugene, with all his apparent limitations, and at the same time experiencing his and Gracie's exceptional gifts had shown him that love and nurturance might be all that is required to make fulfilling lives. Still, he knew that intelligence mattered, and holding onto that, he was conflicted to the point of anguish. He realized that so much of his life's work had been associated with radical social change even to the point of dramatically altering personal lives, even before the effects of nature versus nurture had been fully understood. Yet, the scientific work, and the drudgery of data collection had clearly begun to show the degree to which human characteristics were biologically transmitted between generations. It was as if politics based on assumptions had come to dominate the slog of scientific discovery. And the cost was measured in the loss of human dignity and free will. He had not been in control of any of it, and it all had begun to haunt him.

Rationalizing that she was busy at college, Nigel did not tell Gracie of Julia's imminent departure. In reality he knew that she had always been deeply resentful of Julia's presence in their home, and he feared she might say something that could not be so easily forgiven.

While she was away at college, they at first regularly corresponded, but in time their correspondence became mostly perfunctory, more of the type that satisfies a social obligation. Finally, owing to a particularly heavy workload, she wrote that she did not anticipate returning home for the foreseeable future. Although Nigel continued to write her occasionally, her letters became more infrequent.

Tracing the origin of fulsome resentment can be like digging for a rare archeological find. Layers of meaningless fill would have to be dug away until a first shard of evidence might emerge. Within each subsequent layer, more clues would only hint at what lies beneath. Until sufficient pieces were found and reconstituted, one could not begin to deduce meaning, and even then, one would need context in order to understand their significance.

Perhaps it was talk of his work, about which, from an early age, she sensed something ominous. She was only much later to realize that there was, among Nigel and his colleagues, a seeming compulsion to reorder the human landscape, a veritable plucking away and discarding of those less virtuous among the human family. Then, there was the early rejection of his very own son, who though non-conforming in most measureable human characteristics, was in every sense lovable and capable of expressing love. Finally, there was her father's apparent disregard for norms of loyalty and grief. Having only recently lost his lifetime partner, he formed an abnormal bond with an attractive younger employee, and even brought her into their home. To Gracie, that, perhaps more than all the other hurts, was a betrayal, which compelled a change in how she saw her father. Gracie had never been one to fixate on feelings, but the emotional detritus of family life had accumulated in layers, later to reveal the brokenness of her relationship with her father. In the end, she used the pursuit of a degree, and her plans for continued study as an excuse to begin to pull away.

The day Nigel took Julia to the train station was bittersweet. As the train screeched to a halt, the two who had so long worked together toward great purpose bade each other a solemn goodbye. They swore solemn oaths to write each other regularly. It was their way of assuring each other of a continued relationship, though privately neither believed it. Stepping down from her coach, Julia moved toward Nigel, and hesitated as she looked up into his eyes. Just as she leaned in to kiss his cheek, he awkwardly turned his head. Her lips brushed ever so slightly across his as they shared an embarrassed laugh. Anyone else on the platform might have thought for an instant that she lingered just a little too long.

As they separated, a mail boy came running down the hill to the platform. He called out, "Sir ... Mr. Wellbourne. Dr. Mogridge would like to see you in his office. He said to be sure to tell you, 'right away.'"

"Why, what's the matter?" Wellbourne responded, growing anxious about what must have been a matter of such urgency that it could not have waited until he returned to his office.

"He didn't say. He just asked me to fetch you up there right away," the boy replied. The boy turned to race back to the main building.

Julia climbed the steps to her coach and turned to the platform for one last look. By then, Nigel was already on his way up the hill and did not notice her wave goodbye. Nigel trudged along slowly, hesitating just inside the grand entryway before finally climbing the steps to Mogridge's second floor office. It was his nature to slow down when others around him reacted in haste. It was his way of gaining a measure of control. Besides, it gave him time to think through possible courses of action to whichever might have been the precipitating cause. Whatever the matter, Nigel did not like to be rushed.

Mogridge stood thinking as he looked out of the tall windows facing the central courtyard. From his high vantage point, he could survey the quadrangle formed by the institution's expansive wings. The quad, as they called it, was the central gathering point for many of the institution's ceremonial events. In all but the most inclement weather, it was where the inmates' day began with rigorous exercise followed by announcements. This was Mogridge's ritual, something he insisted on when he took over as head of the institution. It was, he said, among the best ways to unify those who resided there.

"Ah, at last you are here," said the doctor, who exhaled smartly through pursed lips. "There is a matter of some importance. It directly pertains to you." The way he put it, and the long pause afterward had the intended effect. Anyone less socially obtuse might have seen serious trouble brewing right from the start, given the urgency of the call to come to the superintendent's office. "You see, Nigel, we have a problem; that is to say, you and I. It is with your son Eugene. Let me get to the point. Eugene was seen in the cattle barn just yesterday. He was observed kissing Horn's young daughter. I believe she is thirteen."

"I see," said Nigel. "And …?"

"And," he paused. "This is apparently not the first time," Dr. Mogridge added. "I have known of a similar instance for several weeks now, choosing to let it go as some kind of aberration. After this second occurrence, I have taken the liberty of making further inquiries. Miss Mays has also told me, quite confidentially, and to my direct questioning, that Eugene has also been known to pleasure himself."

"Certainly you know, Dr. Mogridge, that with respect to self-indulgence, Eugene is not alone. For I myself have observed such behavior in any number of boys. That behavior is not all that unusual for any adolescent, certainly not for those in our care."

"Yes, but not in the shower … in the presence of other boys," the doctor added. "I'm afraid, Nigel, that we must take steps to address these problems," he continued.

"In the presence of other boys? From whom did you hear such a thing? One of the boys?" he asked incredulously. "Just how, may I ask, do you intend to address these so-called problems?" inquired Wellbourne.

"For now, I think it best that you say nothing of the matter to Eugene, and that for a time you avoid any direct contact with him. That may go far to alleviate any suspicion of favoritism. We will get back to you on just how we intend to address this unfortunate situation."

"I understand," replied a now much-deflated Wellbourne.

"In such situations, it is necessary for me to form a board of inquiry. Ordinarily, you would be involved, but in this case…. well… I'm sure you understand. From there we can make a determination," the doctor concluded.

Dejectedly, Wellbourne left Dr. Mogridge's office. Nigel took the balance of the day off to brood over just what it was that might be done to his son.

Nothing more was said of the matter for the next three weeks. The wait was agonizing for Nigel. Meanwhile, Eugene was kept on his floor during the day to perform housekeeping duties. At the end of a seemingly interminable wait, Nigel was informed that the matter had at last been referred to the State Board of Control, and that they had conceded to a procedure which was intended to inhibit if not completely eliminate any sexual urges on the part of his son. At the Iowa Institution for Feebleminded Children, that usually meant a vasectomy. More recently, however, and under the enthusiastic advocacy of the resident physician from the Kansas School for the Feebleminded, who was then consulting with the state of Iowa, a complete castration was deemed preferential. While vasectomies adequately protected against procreation by defectives, he advised that such a procedure would not inhibit sexual desires. Thereafter began a period of experimentation in which more aggressive measures would be tried. Eugene was to be Case Number 1 at The Iowa Institute for Feebleminded Children.

The procedure had the desired effect. It's just that the operation, and the explanation to Eugene that followed, resulted in a marked change in his personality. Though physically recovered, he was no longer the outgoing person he had once been. He became severely depressed, and his depression resulted in nearly complete withdrawal from all social interaction. Although he would continue to maintain his personal hygiene and could adequately perform his brick-making duties, a position to which he was now limited, he seemed to go through the motions in a near catatonic state. At times, he would simply wander off, only to be found later in some out-of-the-way place on the farm. Nigel was unable to bring him out of it, and following his several entreaties, Gracie agreed to come home for a brief visit. Even she was unable to elevate Eugene's mood.

At the end of the summer, as the nights were beginning to cool and the sumacs took on a glorious crimson shade, Nigel was once again urgently summoned, this time to the big barn across the tracks from the rail shack. Aroused by the fear in the messenger's eyes, Nigel raced down the hill. There he observed a cluster of boys gathered at the base of the large tiled silo, which stood mere steps from the main barn. In the middle of the group lay his son, motionless and muddied. Eugene's head rested near the silo's concrete base. Nigel knelt by his side and gently brushed the mud from the boy's face. The other side of his head was matted with fresh blood. A deep, curved gash ran from just above his right eye across his temple nearly to the top of his head. While Nigel stroked his son's head, the boys stood silently in a circle around the two, mouths agape. Horn's son appeared momentarily from around the side of the barn, then skulked away unobserved. Dr. Mogridge and Dr. Finster, the institution's attending physician, arrived and took charge of the scene. They shooed the boys away. Eugene was pronounced dead, and lifted onto a hay wagon. Horn drove the tractor and wagon up to the infirmary, where the body was hastily examined and quickly prepared for burial.

After a brief inquiry, the cause of death was officially ruled a deep brain trauma resulting from a fall. An 'accident' was how the incident was recorded in the Biennial Report to the State Board of Control. There was never any mention of Eugene's castration in that or any other report. Neither, for that matter, was there ever a formal report to the Board of any other sterilizations at the institution.

All of the evidence would have been puzzling to anyone at the scene who might have been observant. Near the base of the silo, but some distance from Eugene, a long-handled shovel lay on the ground. No one noticed the blood on the upturned blade. Eugene's body was found close by the iron steps that climbed the silo's outer walls. Only Dr. Finster noticed the crescent-shaped gashes in the dirt next to the shovel, but he thought nothing of it. He did not realize that an attempt had been made to clean the blood off the shovel's blade. Neither Finster nor Mogridge paid particular attention to the blood on Eugene's trouser legs, nor the contusions on his calves. They both assumed that the cuts on his leg must have been a result of the fall.

One day after he died, Eugene was buried in an unmarked grave in the institution's graveyard. The graveyard was located on the farthest boundary of the property, near a small grove of trees and brush. A barely visible two-track dirt road led to the burying ground.

Without delay, Nigel let Gracie know of Eugene's sudden passing. Two days after he was buried, Gracie stood next to him by Eugene's grave, weeping. Gracie said a short prayer for her brother, then placed a small bouquet of flowers on the newly disturbed soil. The wind rustled the leaves on the nearby sumacs. Nigel began to sob quietly and said nothing. Gracie turned to look at him, then turned away. They made their way back to the road leading out of the cemetery, and walked home in silence. That evening over supper they spoke only briefly, and that limited to Gracie's academic plans. The next day, she returned to Lincoln.

(Whither leadest thou, oh muse, precious daughter of Zeus? Have you not mislead those learned men of science and letters, thinking men, men who so inspired would direct the affairs of others? Is it not confusion between freedom and submission that leads man asunder? In choosing hubris over humility, and dominion over restraint, are men not then acting as lesser gods?)

[What? What is that you are going on about? You sound a bit deranged.]

(Oh, just talking to myself. Anyway, what exactly was it that you were expecting, given that we have left things alone to work themselves out, as you have insisted? None of this has gone well.)

[Well, you needn't be so snippy. You, of all people ought to recognize that what we are dealing with is creation itself. It is premature to think that what has been created, with such inherently great capacity for making choices, is not capable of making the right ones ... even if it is only one person.]

(Wasn't it that 'great capacity' which gave rise to this very abomination? And of the right choices, can it be said that great progress has

been made, or any progress? It seems like progress is more like a circle than a straight line.)

[My dear man ... your perturbations arise from an intemperate desire to see change happen overnight through some *deus ex machina*, **and failing that, to wield your own intemperate word.]**

(Isn't it also true that your forbearance too often suggests indifference, if not neglect? Can we expect this chapter of man's attempt at progress to be closed at long last?)

[I concede that my seeming lassitude is misunderstood, but do consider our man Wellbourne. Has he not made changes in his life? And are they not for the better? To your last point, would you not agree that people must be given the opportunity to achieve some degree of moral progress on their own? In any event, don't you think it time that you and I should wrap up our little tale?]

(You're right; it is time. As to Nigel Wellbourne's progress, I must admit, he did indeed change. So distraught and overcome with guilt after Eugene's death, and betrayed by ideological zealotry, he turned away from the eugenics movement altogether, abandoning what he had once considered his life's mission. He ran from his past as far and as fast as he could —— to California, where until his later years he served as a remedial math instructor at a boy's parochial school. There, he lived out his final years in self-imposed exile. Save for the time he spent in church and confession, and caring for his beloved schoolboys, he was largely alone. Perhaps the sacrificial life he chose in service was the price he paid to seek redemption.

As for Gracie — she went on to complete her doctorate in animal science, taught at Iowa State University into her 70's, and continued to research and publish scholarly articles on large animal breeding. Although she corresponded for a time with her father, their letters became more formal and less frequent, until they ceased communicating alto-gether.

Harry Laughlin finally retired from the ERO, suffered from epi-lepsy in later life, and unlike other epileptics for whom he deemed it essential, was never institutionalized — a fate marked with no small irony. He returned to his home state of Missouri where he and his wife are now buried.

Now, as to man's moral progress … well, we'll see. Anyway, I think we've made our point, and, at long last it is indeed time to end this story.)

[Not so fast, my dear Consigliore. Our impertinent commentary has served some purpose, in view of humankind's earlier mistakes. That part, at least, ought therefore to end. However, I am alone prepared to continue our story with the expectation that a more hopeful point will be made. And as you well know, the remainder of our story will tie quite nicely to the one we have just shared.]

PART II
ANN AND EUGENE

CHAPTER I

AT THE START OF EACH SEMESTER, Dr. Ann Langley took nearly half the first class period to go over housekeeping items. In what for her had become tedious, she had to explain in detail where to find the syllabus and reading materials, how to turn in assigned work, expectations for class attendance and participation, the testing schedule, and her grading system. She knew students could tolerate no ambiguity in how they were to be judged.

Langley looked over her shoulder at the projection screen, then turned back to scan the mostly blank faces of students taking her class in Behavioral Genetics. She began, "In the seventy-odd years, beginning with Sir Francis Galton's work of the late 1800s to the mid-twentieth century dissipation of the science of human betterment known as eugenics, meaning 'well born,' the movement surged, crested in a tidal wave of progressive social engineering, and in a convulsion of so-called genetic purification left true science in its wake. Ultimately, those with what were considered socially dysfunctional behaviors and physical and mental disabilities were measured, categorized, segregated, neutered, and even exterminated. To identities of class, race, and religion were added categories of behavioral traits to become the essential measurement of humanity." She paused for the weight of her assertions to sink in.

"Culling the unfit, the derelict, the defective, and the degenerate by institutionalization and sterilization was accepted practice. By the mid-nineteen thirties, thirty-seven states had imposed statutory requirements governing prescriptive sterilization. More than 80,000 people in the United States were subjected to its strictures. Compulsory sterilization began slowly in 1907 and then accelerated, following a favorable 1927 Supreme Court ruling in Bell v. Buck. Summarizing for the court, Justice Oliver Wendell Holmes, Jr. said, 'Three generations of imbeciles is enough.' Laws regulating sterilization remained in effect into the 1970s. In time, every state would build physically imposing gothic institutions that isolated the disabled.

"The Birthright Movement, initiated by Margaret Sanger, and supported by the ACLU, the NAACP, and establishment elites, sought to eradicate problems associated with the lower class through positive and negative birth control. Brazil implemented "whitening," a move-ment to breed out what they saw as the nationally limiting consequence of the then nearly predominant Negro race, and the National Socialist movement in Germany committed genocide in the name of racial purity. In America, Better Baby contests morphed into Fitter Families for Future Firesides, and immigration laws imposed stringent limits on the admission of southern and eastern Europeans, particularly Italians and Jews.

"Despite the shock and awe resulting from the extremes of its application during World War II, the eugenics movement persisted, shifting and adapting along with evolving cultural norms. Leading progressives such as Andrew Carnegie, Theodore Roosevelt, Helen Keller, George Bernard Shaw, Margaret Sanger, Winston Churchill, Alexander Graham Bell, and W. E. B. DuBois, to name a few, supported its tenets. The principals of eugenics were well regarded and widely accepted. It was not a fringe movement. Many a small town in America had been exposed to its beneficial claims through news stories or lectures by leading local luminaries, including religious leaders. It is no accident that the given name of Eugene was the most popular name for American boys during a time of widespread acceptance of eugenical principles. That name only began to decline in popularity in the late 1930s, but was seldom given only after the horrors of World War II.

"Eventually, the eugenics movement waned and was thought to have completely faded away as societies around the world became more enlightened. Then, deinstitutionalization of the mentally ill, the epileptic, and the cognitively impaired began. Labels for cognitive disability changed several times, each time sounding more euphemistically humane. States abandoned forms of obligatory sterilization. The one-time rush to use intelligence testing, at least directly, declined as a guide to educational and vocational placement. Society instead began to emphasize the predominance of 'nurture' in human development. In the same vein, laws and the courts changed as well, seeking to recognize the dignity, worth, and rights of the human born. Ultimately, the tenets of the eugenics movement were rejected as pseudo-science, if not maliciously fraudulent, and work on inherited human differences would ultimately, at least for a time, fall out of favor. Let me just add one final comment. In retrospect, the smugness, the hubris, and the very audacity of this progressive social engineering is breathtaking. Today, we may think the idea of altering a population's mating and reproductive choices is delusional. Yet, we have had China's one child policy, Paul Erlich's 'Zero Population Growth' ... and more. All this in the name of science."

Glancing at her notes, she paused to assess the impact of these opening remarks on her students. Most were dumbstruck. Many had never heard of eugenics, at least beyond the horrors of Nazi Germany, and even about that, many were mostly ignorant. A few who thought they'd maybe once heard the term confused it with Scientology and its system of Dianetics.

Looking at her students, now with their heads down and their hands in their laps, it wasn't clear to Professor Langley if they were too preoccupied with their phones to hear anything she said. No matter, she thought. It became her practice early in her teaching to absolve herself of responsibility for motivating students, that is, beyond the considerable effort she put into making her lectures and class assignments engaging. She thought to herself that it would be no different this time. In the fifteen years she'd been teaching, students hadn't ever been too different, at least insofar as dedication to their studies went. A select few students would be so deeply engrossed in the subject matter that it might influence the future direction of their education, and perhaps even the direction of their careers. These were the students who would learn to care passionately about making advancements in human biology. They were the ones in which she took the greatest personal interest, and they were among the very few who might be encouraged to consider pursuit of a terminal degree in her discipline. If chosen, they might become her acolytes and provide the assistance Langley would need to pursue her own research interests. For the rest, though, while the subject matter might stay with them, at least for a time, the course was at best merely the fulfillment of some credit requirement necessary for graduation. She accepted her responsibility to them all by delivering a class that easily met the necessary academic requirements while providing sufficient challenge to test even the most able student. In contrast, many of the graduate students in her class, being high need-achievers, would be more focused on merely earning a good grade. Whatever the level of commitment of her students,

Professor Langley was consistently rated as a five-star professor, and her course was always highly recommended.

"In this course, we'll not only review the history of research into heritable behavioral differences, but also explore the present state of knowledge of the impact of genetics on all manner of human behavior. At the completion of this course, it should become abundantly clear that aspects of all behaviors are heritable to some extent. For those of you who may have long clung to the idea of the indomitable power of nurture, and the brain's so-called plasticity, the very idea that we are mostly a tabula rasa, a veritable lump of clay to be molded by human experience and instruction, this material may come as a complete shock, if not heresy. So fasten your seatbelts.

"Now, in the syllabus you'll see a link to Thursday's reading assignment, 'A Brief History of the Eugenics Movement.' Please come to class prepared to discuss this material. Pay particular attention to the study questions, which will be the foundation for our discussion. Are there any questions?"

A handsome, intense-looking student in the front row was first to raise his hand. He appeared slightly older than most in the class. His rugged appearance was arranged with studied care, which conveyed both masculine strength and attention to detail. His hair was carefully tousled. His slightly heavy one-day-old beard would look the same every day during the semester, and his branded outdoor sportswear, though well-worn, was never the least bit unclean or out of date.

"Yes," she responded, directing her gaze to the young man, whose hawk-like gaze suggested a degree of focus that might have been unsettling to one less self-possessed than Professor Langley.

"It's apparent in your introductory remarks — for example, you said, '…was thought to have completely faded away,' and you referenced shifting cultural norms, and alluded to 'direct measures of intelligence' — does this mean that you believe that the eugenics movement is still alive?"

Professor Langley fixed her eyes on him more intently. In a millisecond she posited that he was likely among the most incisive among this class of students. She looked down at her seating chart to get his name, noticed that he was a grad student, then connected his name to her impression of him — 'Hawk.' It was a mnemonic device, a habit she developed as an assistant professor in order to connect a name with a face. Langley was one of the rare college professors who used a seating chart, at least until she got to know her students' names, never mind that she might never see most of them again.

Staring back at Hawk, she thought that she might make it a point to find out who his academic advisor was. Then she said, "I will leave that for you to conclude ... um, Mr. Augustson." She turned away from him to see if there were other questions. "Yes, the young lady in the back row," looking down then back up, "ahh ... Miss Jensen."

"Dr. Langley, you mentioned positive and negative birth control. I gather you are referring to abortion in the latter case. Don't you think it is advisable to provide some kind of warning in the course description, and certainly before starting your lecture, if you are going to be covering such controversial subjects?"

The professor had known such a comment would come. They always did sometime over the semester. For more than five years, she could count on some of the students to express special sensitivity, if not outrage, at her material. Invariably, it would be on those topics that could offend the most delicate sensibilities. "I'm sorry, Miss Jensen. I had thought that the course description made clear that controversial topics would be covered. Perhaps you should consider dropping this class for something more suitable." Nervous titters could be heard as heads swiveled to observe the student in the back row who had asked the question. Miss Jensen's faced burned with embarrassment, as she attempted to hide her humiliation by reaching down and fumbling with something in her backpack. Langley realized that her response was harsh. For some reason she lost control this time. Ann had been feeling more tension in her classes of late. Trigger warnings, safe zones, speech codes, protecting feelings and other such social imperatives had worn so thin that she could no longer brook the now well accepted norms of her institution. By the time she reached forty-one, social convention, at least on a college campus, had lost some of its standing. Besides, she reasoned, her prominence in the field of behavioral genetics had given her sufficient stature to stand against institutional and cultural tyranny — of course, tenure played a part. She no longer need fear the social police, believing that at worst she might merely be ostracized by her peers, or counseled by the administration. Such annoyances mattered little to her. She knew that her work could still be carried on independent of administrative interference.

Even as a young girl, and especially as a woman in science, Ann resolved that fear would not be an impediment to achieving success. She never sought notoriety, though she accepted that it could be helpful. Rather, it was the thrill of making some type of breakthrough discovery; that is what drove her. Unlocking just one of the mysteries of human biology, particularly if it was foundational, enabled her to endure the rigors and the sometimes-joyless burdens of teaching. Despite the many indifferent and modestly-abled undergraduates, a stultifying bureaucracy, and the perpetual need to solicit research funding, the work was still mostly fulfilling.

Before the nervous laughter at Miss Jensen's expense had died down, a wisecracking student sitting by the door asked of no one in particular, and in seeming seriousness, "Eugenics — is that like another word for eunuchs?"

Heads swiveled in his direction, aghast at the stupidity of his remark.

"No, stupid," his seatmate replied. "Eunuchs are castrated males. It's what they ought to do with some of the football players." Of course, everyone understood the reference, given that a star athlete had recently been suspended for some sort of undisclosed sexual impropriety.

Uncomfortable laughter rippled through the class, a reaction born more of awe than of delight at the exchange. While Professor Langley sometimes admired the irreverence of youth, and nearly always the occasional, astutely observant flippancy, she held that reaction in check this time, dismayed at the manifest lack of self-awareness exhibited by these two students.

Langley thought that higher education had once been a societal winnowing process. Maybe it still was, but to a lesser extent. Admission policies were once far less egalitarian. Many secondary students once would not even have gained acceptance at a four-year university, owing to limited financial resources, manifest cognitive limitations, or both. Save for legacies, children of major donors, and athletes, such impairments were well signaled by the social standing of one's parents. The decision to deny admission was an unashamed conclusion that affected students would be poor prospects for higher-level academics and timely graduation. Now standards were lower, consistent with a broad expectation that nearly everyone graduating from high school must be college-ready. For a growing number of institutions, submission of college admission test scores had even become voluntary. Easily accessed government-backed student loans, without any form of underwriting, would make financing possible and broaden the range of academic subjects one might pursue to include those of no discernible economic or social value. As one might predict, students of lesser ability inevitably would drop out after the first or second year, it becoming apparent to nearly everyone, including themselves, that the intellectual and financial rigors of higher education would require more than they or their families could justify. This phenomenon had the most disadvantageous effect on those who selected the least vocationally oriented academic fields, and in Ann's view, the least rigorous majors. Moreover, she thought, the smorgasbord of classes, even those that might comprise a major, were of such limited utility that any learning would be soon enough forgotten. The labor market did not value achievement in these courses. What it valued was the presumed intellect and tenacity that was signaled by completion of a four-year degree. Professor Langley felt that college should be, as it once was, reserved for the most intellectually advantaged. In her view, tests of general cognitive ability, given at an early age, would have been a far more efficient and equitable way to achieve the goal of sorting for the kind and level of education, and concomitantly the work for which one was best suited. A more utilitarian and vocational emphasis in education, one which was tailored to all levels of intellectual capability, would have made for a vastly more sensible and economically efficient system from grade school through post-secondary education.

"Before we adjourn," she redirected, "are there any final questions? Yes?"

"Will there be a required textbook?" a student piped up from a far corner of the class.

She didn't bother to look up his name. "Perhaps you missed the start of class and my opening announcements; or perhaps you've overlooked what was in the syllabus when you enrolled. As you should

have noted, we will be using a variety of materials, available online …
at no cost to you. So, no, you will not be required to buy a textbook.
And yes, there will be a midterm and a final exam. If there are no more
questions, we will see you Thursday." She gathered her belongings and
left for her office upstairs.

◆

Langley walked into the classroom the following Thursday,
girding herself for the moments of heat if not light that she knew would
come in this particular class period. There had always been much of
the former during the second day of class. In fact, some students could
be counted on to completely lose self-control during this segment of
the course. Mention Hitler or the Nazis and someone could be counted
on to get overly agitated, perhaps even reduced to tears. So, as a matter
of necessity, she always began this class discussion by going over
ground rules: show respect for others, especially when disagreeing;
support assertions with facts; and stick to the topic. Do not make it
personal.

"Based on your reading, who would like to make a general observation about the early period of eugenics history? Yes, Mr. Herndon
… Charles," Langley gestured in his direction.

"It seems like Galton and others were trying to shift from just
describing things to actually measuring them," Charles started.

"How so?" asked Langley.

"Until Galton came along, and later Pearson, scientists mostly
described phenomenon by using sentences, adjectives, and all like that.
Biological science was mostly based on morphology. I think that's the
word. Later, using statistics, they began to measure things," Charles
answered.

"What kind of statistics?" she asked.

"I guess it was frequencies, and averages, and, oh, yeah, later, correlations."

"Things? And what were some of those 'things' they measured?" she asked, emphasizing her distaste for his imprecise word choice.

"Oh, like physical deformities, the number of those in a given family tree. Also, mental illness, and drunkenness, too. For example, they began to correlate the closeness of family relationships with these things."

"Okay," Langley said as she rolled her eyes. "Any problems with what they did? Yes, Mr. Chen."

"Once they started correlating things … er, I mean, physical and mental characteristics, it seems like they went too far," Chen answered quickly.

"In what way?" Langley queried. "Can you give me an example?"

"Yeah, sure. So, for example, like they said that TB was caused by poverty," he replied.

"Good," she said. "They weren't too far along at that time in discovering an underlying viral vector for TB. Who else has a general comment about the early eugenics movement?"

The Warrior, as Langley came to think of Miss Jensen, now standing in the back far left corner of the class, spoke up, her voice quavering with palpable anger. "The science of eugenics, if you want to call it that, was racist, xenophobic, homophobic, and ableist … and patriarchal!" A small spray of spittle could be seen as she accented the list with pejorative inflection.

The class pivoted to see whose words dripped with such caustic invective. Those now much over-used words were abundantly familiar and arousing to just about everyone who read the news, attended a college rally, or had a course in the humanities or the social sciences ... whatever their political leanings. Langley attempted to redirect by asking the young lady for evidence to back her assertion.

She hesitated, then said, "Well, the movement to sterilize blacks, gays, the differently-abled, and mostly young women were certainly examples. Also, immigration laws kept out Jews."

"You're right about all of that, but those practices, those policies came later. Sterilization largely took place in the teens and on into the thirties, when the eugenics movement was in full flower," Langley corrected. "Eugenics had perhaps more noble aims at the outset. Though it might likely be said that Galton may have had his biases, the path of his research was based on improvements in the scientific method, and had as its initial goal betterment of the lot of a large segment of society. This was not unreasonable in a rapidly industrializing, urban environment. Still, I think your points are valid."

A young man who wore his ball cap backwards, the one whose friend made the stupid comment about eunuchs during the first class, turned back to look at Miss Jensen. He then turned to look at his friend. Hat Boy, as Langley would come to think of him, was about to mutter something snide to his friend, she was sure. But his jibe was loud enough for everyone to hear. "Worn out labels have their place ... on the clothes rack at Goodwill." Realizing that he'd gone too far, he turned to face Professor Langley, then added, "For scientific inquiry, though, er, they ... ah, labels I mean, won't get us very far. I believe that was Galton's point. That's why he set out to actually measure human variation."

"You need to shut your mouth, homeboy," the snarly young women commanded, glaring at the back of his head.

The boy with the hat turned and faced his accuser. Smiling impishly, he asked, "Does this mean we can't still go out Friday night?" The young men in class laughed. His friend laughed the hardest. A few of the young women giggled nervously.

Still standing, the angry young woman flipped them off with both hands, then sat down. Minor bedlam ensued.

Ann closed her eyes for a moment. When she opened them, the verbal chaos remained. She walked calmly over toward the door and flipped off the lights. The room went silent. She turned the lights on. "Let's stop right here," Professor Langley barked. "I remind you all that we are going to have a civil discussion, and hear out other points of view, no matter our own personal beliefs. If you can't do that, I'll have you permanently removed from this class. Are we clear?"

To get them back on track, the professor asked, "How did eugenics change over time? Anyone."

There was a protracted silence, as students were now hesitant to speak after the professor's sharp rebuke. Finally Hawk, the intense student in the front row, spoke up. "Even as the science, then still in its infancy, was advancing, the movement seemed to shift in a new direction. By the time the Americans at Cold Spring Harbor, New York, got involved — through Charles Davenport and Harry Laughlin — it became more of an activist social movement. That was decades before the mechanisms of genetic inheritance were very well understood. That's where it shifted from science to activism, and on a global scale. For example, Laughlin was well known by, and especially influential with, the Germans. He was even awarded an honorary medical degree by the Nazis for his contributions to the field. Eugenical science was spreading worldwide."

"And would you say," looking down at her seating chart to confirm the name, "David, that as many have characterized it, the movement was based on pseudo-science?"

"I would not," David declared, "despite the many now apparent errors in their nascent science-based conclusions." From the shifting in their seats, it was clear that most of the students in the class were not ready to accept his claim.

"Go on," Langley encouraged.

"Well, it is apparent now that many of their early observations have since been borne out, at least in part. We now know, though we don't yet fully understand the underlying pathways, that mental illness, and specifically schizophrenia does have a heritable link. This is true with many other disabilities. This was apparent from their work compiling family pedigrees. Similarly, they observed and bore out with statistics that academically successful parents tended to have academically successful offspring, suggesting that general cognitive ability may largely be heritable. Early twin studies seemed to affirm that observation. Though their conclusions, such as the one mentioned earlier about TB, sometimes failed to consider other factors, and attributed too much to heritability and not enough to environmental factors, they clearly were onto something."

"Excellent comments," she said. "It seems that you have gone beyond the required reading, but good! That's good. Now can anyone else think of scientific conclusions that were once widely accepted but that have since been proven wrong, or are at least partially mistaken? Any other scientific blunders you can think of? Any science that might have prematurely shifted to political activism? Anyone … just speak up."

"The earth is flat."

"The sun revolves around the earth."

"Anything more recent?"

Hat Boy's friend couldn't resist. "Fried eggs, bacon, vaping, and … um … ah … dope is bad for you." More uproar followed.

When the laughter subsided, Professsor Langley added with a bemused smirk at this most recent comment, "Who can resist bacon? Now, anything else?"

"Race …" came a voice from the back row. Now standing, Miss Jensen said, "… is a social construct, not a biological reality. There is no real thing called race. It is just a means of systemic oppression."

"Thank you for that observation," Langley responded. "It is Miss Jensen, isn't it?" she asked, looking down at her seating chart to confirm the name. "Anyway, Miss Jensen, this view originally harks back to Franz Boas, a noted anthropologist. Still, most people see race as biologically determined. Perhaps your assertion stems from something you have learned here at this very university, but in this case it is not completely accurate, at least the part about the biology of race. The truth is a bit more complicated." She paused to take a deep breath and gather her thoughts. "In fact there is indeed some genetic basis for racial categories. Let me explain. In the human genome, there are on the order of three billion base pairs in the double helix you now recognize as DNA. Base pairs are made up of the proteins adenine, thymine, cytosine, and guanine, known by letters A, T, C, and G. Base pairs form individual rungs in the spiral ladder that make up the DNA strands. Long sequences of these base pairs in segments of DNA make genes. Genes code for proteins in cells, and those proteins affect cellular function. Individually and collectively, they affect human characteristics, at least in part. The human genome has around 20,000-25,000 protein coding genes. Different genes can be made up of anywhere from 27,000 to 2,000,000 base pairs. As I said earlier, the total number of base pairs in the human genome is around 3.2 billion, so that leaves a considerable amount of genetic material that is not considered a gene. More on that at a later time. Anyway, the complete genome is made up of around 6.4 billion individual bases, all divided up and tightly bundled in 23 pairs of chromosomes in every human cell. So you can see the numbers get quite large, but the structure of DNA itself is very, very small. Are you with me so far?" She paused to see that all of the students were now staring intently at her. She went on, "While we like to say that there is only a 1% difference between the genomes of humans and chimpanzees, that 1% amounts to 30 million base pairs … which is quite a large difference after all. I think you can agree, we are manifestly different from primates … mostly. Anyway, I think you would also agree that the difference between us and primates, and other species for that matter, is biological, and not a social construct." She smiled as she glanced quickly at Hat Boy's friend. Hawk inferred the meaning of her barely noticeable glance. "Now, all humans basically share the same genes. So does that mean there are no different races, that we're really all alike, or even very much like monkeys? Well, we know from our own general observations that this is not true. Stay with me. First, for any given gene that humans have in common, two individuals may have slight differences in the letters, the base pairs that make up that gene. Some would say that among humans there is only a .1% genetic difference, but .1% of 6-plus billion bases still leaves a substantial number of differences. Those differences we call alleles, or different forms of a gene. Yet it's still the same gene, and those alleles may differ in only a few bases, or in thousands of bases. Sometimes the differences don't seem to affect observable characteristics, or traits. Those characteristics we call phenotypes. But sometimes they do. I see many of you trying to write all this down. Don't bother just now; we'll come back to it. Anyway, where was I?

"Human evolution, which is continuous, has created clusters of alleles, or clusters of genetically varied populations, particularly in groups separated geographically or culturally. This is especially true if those populations are or have been isolated, at least for long periods. The frequency pattern of alleles in those clusters results in different phenotypes, or different observable characteristics. Clusters of certain phenotypes in our example can be called race, which is for some a useful construct that distinguishes observable traits among various people groups, which is a better term. We have given familiar names to these people groups: Caucasian, Sub-Saharan African, Asian, and so forth. Phenotypical variations of race include physical attributes — skin and hair color, for example, nose shape, nutritional metabolism, brain function, disease process. Let me be clear, though, these characteristics can and do change over time for a variety of reasons. Within a particular racial category there can be wide variance. For example, there is as much or more genetic variation within Africans, within what most have come to simplistically think of as one race, than there is between any two other races. Since all people groups represent considerable blending with other people groups over very long periods, we are each an amalgam. So yes, it is true, all people have much in common with each other, and even with chimps." She glanced again at Hat Boy as she said this. "But still, there are important differences, some of which we'll get into later in the class.

"Let me also add that aggregating clusters of genotypes into a construct like race is not particularly interesting to the behavioral geneticist for anything other than looking at population-based associations between genotype and phenotype. The medical establishment, cultural anthropologists, social scientists, and public policy makers may have interest, but we generally do not. Analyzing other phenotypical clusters is of greater interest. For example, we might look at clusters of peoples with greater average life spans for associated genetic variance. Or we might look at differences in general cognitive ability for different subgroups … say Ashkenazi Jews. The goal for the geneticist is not deconstructing social power structures.

"To summarize: race can best be thought of as a generalization, and not always a particularly useful one at that. As with studies in virtually every field, phenomena are grouped into categories for purposes of study, though there will always be variability within and generalization about any category. As you will see, much of what we do in science is investigate generalizations."

Miss Jensen, still standing, shouted, "That's racist. This class is racist, classist, xenophobic, and no doubt homophobic and transphobic, too! You're preaching hate, not teaching science." She hefted a considerable backpack onto narrow shoulders, and then marched out of the room, pausing long enough at the door to shout, "I intend to report you to the administration."

Her outburst incited a more generalized disruption that gathered momentum as students commented among themselves about what they had just witnessed. Before it had gone on too long, Professor Langley took control.

"As you can see, the topic of eugenics commonly elicits a passionate response. It is a movement that, though perhaps begun with good intent, got caught up quickly in social policy and moral values that today can still evoke outrage. Lest we think this is a problem only for an earlier time, let me just ask, is there any other scientific movement of our time that has shifted from science to social activism, perhaps before we have fully understood that science, or adequately wrestled with associated moral and social issues?"

Without waiting to be called on, David spoke up. "Yes, climate change. One cannot help but see that there is fervency behind the climate science movement that has become less about advancing the science and more about politically directing substantial changes in the way we live. In particular, advocates would require an abrupt shift away from fossil fuels, never mind the economic impacts and attendant social costs, which I might add are more likely to be felt by the poor. Yet all the climate models on which the movement relies fail when back-tested. Of course, that doesn't mean there isn't anthropogenic climate change. It simply means that the possible response has maybe gotten ahead of the science."

Langley offered no comment to his observations, anticipating that pursuing this topic would likely distract her students from the topic at hand. "Thank you. I think we need to move on.

"So then, why did the eugenics movement apparently die out? Let's hear from someone else." She looked at her seating chart and called on Marisa.

"I suppose it was the horrors of Nazi Germany, and their attempt to achieve racial purity that so shocked the world and embarrassed the thought leaders of the movement. The science was in conflict with a common sense of morality. That caused it to die swiftly, I think."

"And do you suppose that the early aims and methods of eugenics are in fact completely dead? You don't need to answer that just yet. It is something that I wish you to consider over the course of this class," the professor said. "Anyway, next Tuesday we'll dive more directly into behavioral genetics. We'll look at the nature vs. nurture questions in two case studies, first by defining behavioral genetics, then examine the John and Joan case, and finally a genetic condition known as phenylketonuria, or PKU. You'll find the reading assignment and links to two short videos in your online syllabus. Oh, and one last reminder, I have open office hours immediately following class on Thursdays only. At other times, you will need to make an appointment."

◆

Leaving her office door ajar, Langley set her large shoulder bag on the floor by her desk and plugged in her laptop. She dropped into her chair, kicked her shoes off and exhaled deeply. Given the way in which the class discussion had proceeded, or rather exploded, she reflected that it was going to be a long semester. For a moment, she was perturbed with herself that she had let things get out of control, but then she reminded herself that she never really was in control. These were college kids, and their attitudes were different today. Social boundaries were breeched long ago. Maybe the best she could hope for was to referee, to make sure no one got really hurt.

Lost in thought, she did not hear her friend Lilith step into her office. Lilith Hunter, or Dr. Hunter as she insisted for everyone else, was a professor of philosophy at the university. Her office was in an adjacent building, and her teaching schedule coincided with Langley's so they were often able to have lunch together. Save for teaching, lunch, or open office hours, Professor Langley was usually in the lab with her post docs and the few graduate students she employed. Given that the semester had just started, Dr. Hunter was reasonably confident that Langley would be in her office tending to administrative matters at this time of the afternoon.

"So," Lilith asked, "how did it go?"

"Hey, I didn't hear you come in. Oh, god; I've got a good one on my hands this time. I'm a bit ashamed of myself for letting it get out of hand. A bit off my game, I guess," Langley answered.

"How so?" inquired Lilith, eager to hear of someone else's misery. Dr. Hunter was always keen on hearing any tale that might break the tedium of her own rather soporific subject matter, and knew that controversy was inevitable when eugenics was the topic of discussion. For the better part of fifteen years, especially at the start of a new school year, Lilith had pondered how it was she ever decided to pursue philosophy as a subject. It must have been some late night undergraduate bull session, or the effect of one or another of her poor choices, namely a mind-altering substance. Lilith's reflections tended to be like that: hypercritical or self-disparaging, or worse, both.

"A couple of my students went at it. Nothing physical, but one walked out in a huff. Don't think I'll see her again. Just as well. She was a walking social justice poster. Turns out her provocateur was a clever enough boy, in spite of an appearance which suggested a severe case of arrested development. I did appreciate his wit, though. But in combination with the young woman, it was a combustible mix. To add to these two, we have the clever boy's dull-witted friend, whose inappropriate comments, at least for a college classroom, only added an accelerant to the fire."

"Emotional incontinence. So many have it today. Anyway, it all sounds delightful. Can I come and watch on Tuesday?" Lilith quipped.

"You need a new career," Langley retorted. "Anyway, it's remarkable how different classes are from semester to semester. I've had controversy before, but never one with such a … uh, a snippy outburst. Little things can go so terribly wrong. Just like SNPs I guess."

"Jargon again?" Hunter asked. "What the hell are SNPs?"

"Single nucleotide polymorphisms — little variations in the genetic code, like letters out of normal sequence. Actually, they're quite normal, occurring 1% or more of the time. Given that there are more than 6 billion letters in our genetic code, errors are quite common. All humans have muddled letters — in other words, we're all walking typos. Our genetic system is made to produce variation — some good and some maybe not so good, but it's all part of the overall design."

"Like so many of the student papers I see," quipped Dr. Hunter.

"Cute. Anyway, I just happened to have a few of them in my last class, walking typos, I mean. And it's more accurate to say the three miscreants I mentioned are more like mutations than SNPs. Quite rare, occurring less than 1% of the time, they are sometimes harmless, but then again, they can also sometimes cause great problems."

"There you go again. Do you ever stop?" Lilith asked. It was more an observation than a question. "Anyway, what I dropped by for was to see if you had any plans for Friday. If not, I thought we might grab drinks and dinner at the Dakota. Eva Johnson is singing, and I've heard you say you like her music. What do you say?"

Just as she answered, Hat Boy knocked at the partially opened door, then stuck his head in. "Did I catch you at a bad time, Professor Langley?" he asked politely.

"My open office hours are every Thursday after your class, as I said. So, do please come in," she offered in a pleasant, and reassuring manner.

Turning to find a chair, he noticed Dr. Hunter, who was sitting just out of his view. "Oh, I'm sorry, I didn't realize you were with some-one else. I can come back," he said.

"That's all right, I was just leaving," Hunter announced. "I'll pick you up at six," she said to Professor Langley. Turning to get a good look at the boy, then back to Langley, she smiled wryly and said, "Good luck."

"Anyway, what is it I can help you with, umm, forgive me if I don't recall your name." the professor said.

"Charles, er, Chuck Nathan. No matter, I just came to apologize for my comments in class. I really didn't mean to cause anyone upset. Sometimes without thinking I just say what's on my mind. I guess I sometimes forget where I am. Anyway, I'm sorry, and won't let it happen again," he added.

"You're forgiven, and thank you for the apology. As I said at the outset, the topic touches some controversial areas. It's imperative that we keep some semblance of order."

"I understand," he said. "It's just those words that girl spoke … I get kinda sick of hearing 'em, you know what I mean?"

"I do," the professor said, " but I also know that in this case the young women's comments were accurate, if inflammatory. I'm sure if you did the reading, which it appears you did, that you understand that," Langley said. She was beginning to modify her first impression of the boy, charmed at least by his openness and humility if not the manner in which he wore his hat. "You know, the point of where we were going is that humility, in contrast with hubris, is advisable in the pursuit of science — in anything for that matter." Looking at her watch, she hastened to add, "I really must be going soon. I've a meeting across campus. Anyway, all is forgiven. We'll see you Tuesday, and thank you for stopping by, Mr. Nathan."

Just as Nathan opened the door wider to depart, David, the sharp student who sat in the front row of Langley's class, appeared in the doorway. He stepped to the side to let Nathan exit, then asked if Professor Langley had a few minutes.

"Only just. I've got to leave very shortly. What's on your mind … David, is it?"

"I just wanted to set up a meeting to discuss my academic program," he said.

"You have an advisor, I trust," she replied, sounding a little impatient.

"I do," he answered, "but this is a matter about which I think you can best advise me," he responded.

"Academic program?" she queried.

"Yes," he said. "I'm in my last year of grad school … Masters psychology … statistics and IT emphasis, and …"

"You're at the right school for all that," she said. "You were starting to say something else?"

"I'm not sure what's next. I mean after," he said.

"It's a bit late for uncertainty, isn't it?" Langley asked.

"Not too late, I hope. Look, I know you have someplace else to be, but I wonder if you could give me twenty minutes sometime so that I can get your opinion. That's all," he said.

She had already opened her calendar while he was talking. "Three o'clock on Tuesday," she said. "Here. Will that work?" Her reply was swift and, though abrupt, was not meant to sound discourteous.

"Thank you," he replied. "I'll see you at three o'clock Tuesday."

Just as Professor Langley was preparing to lock her office to head to her graduate seminar, the janitor who cleaned the floor of her building appeared over her shoulder. He smiled broadly, his eyes squinting behind thick glasses. His too-large-for-his-mouth tongue produced a lisp as he greeted her with "Hi, Mithuth Langley!"

"Oh, hi, Mark. You surprised me. I suppose you wanted to grab my wastebasket."

"Yeth, Mithuth Langley." He pushed his glasses up on his small nose.

"Go ahead, grab it quick, Mark. I've got to go," she urged.

When he finished, she locked the door, and headed off to her meeting. She felt a sense of relief. Lilith's visits were always fun, even when she was in one of her moods. Their plans for the weekend gave her something to look forward to. Perhaps, she thought, it was also Mark; he always buoyed her spirits. His simple joy had that effect on nearly everyone. It might have also been the encounter with Hat Boy, whose humble apology was so welcome. Then again, it might have been getting to speak with any of her brightest students outside of class, in this case, David. There was something appealing about him. Funny, she thought, you could take a group the size of an average classroom, throw in a few more to make a decent sample, and you'd have a nearly normal distribution on just about any human dimension.

CHAPTER II

Lᴉʟɪᴛʜ ᴡᴀs ʟᴀᴛᴇ ᴀs ᴜsᴜᴀʟ — not disruptively so, just enough to remind Ann of the minor chaos that enveloped her friend like an aura. As recompense, Lilith wielded wit that was, as circumstances required, capable of eviscerating its victim like a broadsword. Whatever her other peeves, it was pretentiousness that Lilith could least abide, and required of her a just response. In some circles, this was seen as a flaw, but never with Ann. Lilith's incisive comments revealed a keener understanding of the world and its contemporary ethos. Ann liked to think that she shared the attribute, though was at best a weaker understudy. Both could assess humanity's foibles with enough detachment to diagnose and offer prescriptive remediation. Although both shared some degree of affinity for such encounters, it was Lilith who seemed to be energized by them. Seeing the world as it is, rather than imagining it perfected, was one of their main sources of mutual attraction.

Ann and Lilith were no run-of-the-mill iconoclasts. There were already plenty of those on a college campus. As Lilith liked to say, "You don't need nipple rings and a tattoo to be a free thinker."

Their objectively rational view of the world easily put them at odds with many fellow academics, especially those who slighted reason in favor of feelings. With prescribed haughtiness, the academic majority moved through broader policy discussions with studied synchronicity. Truth, academic truth, at least among their peers, was axiomatic and self-evident. Debate was not required except as a pro forma exercise. Mostly, it was to signal collegiality, or to offer abundant airtime to those who had need, and nearly all had need. It was almost never possible to have real debate, because that would have required expressing opposing points of view. On those rare occasions when real disagreement was quickening, debate would be aborted even before the crown of discord emerged. It was not so such much a lack of intellectual integrity or cowardice that inhibited Ann and Lilith from engaging in opposition. It was practical recognition that their positions were lost before they could even be taken up. They knew it was fruitless to proceed. Debate could at worst lead to minor erosion of their independence, or perhaps to censure by those with whom they preferred no intimate association. More to the point, most of the policy discussions were on matters of such little importance that they need not have been taken up at all.

Try as they might to avoid being drawn into any discussion on broader public policy or politics, abstention was seldom completely possible, for even the slightest grimace or silence itself was a tell. So it had become their practice to excuse themselves whenever such occasions might arise, and instead devote themselves to their work. In Lilith's case, it was also the responsibility of caring for her aged mother that allowed escape. On this last point, Ann had once remarked to her friend that she, too, should have such an excuse to bow out of social engagements, anything beyond the briefest encounters with her colleagues. As she once said to Lilith, perhaps she should adopt a disabled orphan, or volunteer at a women's shelter so that she might also have a socially acceptable excuse for avoiding meaningful social interaction with other faculty.

"You look nice," Lilith gushed as Ann got in the auto.

"Thank you. It's left over from before. It was a gift from my ex. I'd have donated it, but something about it I still like. It makes me feel so desirable," she said in self-mockery.

"I don't know why you'd say it that way. You are still ... I mean you are desirable," Lilith said, as she jerked the wheel of her Buick hard to avoid being hit while pulling away from the curb. "Asshole!" Glancing over at Ann with a smirk, "Not you," she added. "Jerk drivers. Too busy playing with their dingalings to let a girl go. I signaled, for cripes' sake."

Lilith's driving was as erratic as her life. Ann always said that you could tell a lot about a person by the automobile they drive and how they drive it, an observation that admittedly was redundant.

"Anyway," Lilith said, "I'm looking forward to hearing Eva. I haven't seen her since we went to the jazz festival last year. Have you?"

"I hear her on public radio, but haven't seen her in person. Speaking of 'haven't seen' anyone, have you heard anything from that guy you were seeing last month?" Ann asked.

"God, no," she said, glancing over at Ann to reveal her disgust. "Guy turned out to be a real creep. More interested in working out with his buddies and showing me his abs than having a meaningful relationship. He did have nice pecs, though. Shame." Lilith finished with a swooning sigh for effect.

"I know the type," Ann responded. "Seems like everyone's into bodily perfection these days. I suppose it is a form of self-expression, but it sure lacks creativity. I guess it's easier than improving the mind."

"Narcissism is what it is. That and fear of relationships," Lilith corrected. "Suggests to me a lack of depth. That's okay if all you're looking for is a physical relationship, but at my age that won't cut it," she said as she pulled into a parking space behind the jazz club. "You can tell me about your love life after we get in and get a drink. You've got more to show for your life experience than I do."

"Couple of ol' cougars on the prowl now," Ann added, somewhat embarrassed at her own utterance. "Not much to tell, but you know me, I hide nothing from my friends."

A swarthy waiter with a man-bun and earrings stood over them, pencil stuck in his hair, a sultry smile plastered across his face. "You up for appetizers?" Ann asked. "I'm not that hungry. You want to split an assortment? You decide."

"Sure, why not? I'm really more interested in a whiskey, but heck, why not?" Lilith replied. Looking at his name tag, "André, the lady and I would like the calamari, avocado dip, and … you okay with quesa-dillas? …. Quesadillas, then. Thank you, André," she finished, looking up under her eyelash extensions, a coquettish smile on her face.

"And for you, miss, something to drink?" he asked turning to Ann.

"Boodles and tonic, please. Thank you," she said. He was not much more than half her age, so it felt awkward being addressed as miss. Her students could never have seen her that way.

"Well," Lilith began conspiratorially, "so tell me."

"Tell you what?" Ann replied trying to sound obtuse. "Oh, that. Well, there's not much to tell."

"Begin, then, please," Lilith commanded, sounding more like the chair of a doctoral committee.

"We've dated a few times. No spark there. I've tried to avoid him, frankly."

"Oh, and why is that?" Lilith said, now annoyed.

"It didn't take long. It never does at our age. He made it clear early on that he was interested in more than a casual relationship. That's not what I had in mind." Pausing, she added, "At least for now.'"

"What is it with these guys? Can't a girl just be in it for some fun and companionship?" Lilith asked as if she were talking to herself.

"Look, I can't say I wouldn't ever, but for now, my work's too important to get bogged down, and like I said, there wasn't much spark there," Ann finished, just as their drinks arrived.

"Anyway," Ann went on after sipping her gin and tonic, "to be honest, I could actually see myself with a child one day. Not now, but one day, perhaps. I know that sounds silly, but …."

"Silly?!? What the hell you thinking of, girl! You're forty-what … three?" Lilith blustered.

Looking a bit put off, but just for show, Ann responded, "I'm only forty … okay, forty-one, thank you very much. My body is still capable. Been running again, you know," she added, flexing her arms.

"But a baby?" Lilith continued. "Aren't you a little … er, busy for that?"

"Look around, Lilith. Many of our colleagues, at least the tenured ones, even some on the tenure track are married with children. There's no reason I can't," she answered defensively.

"None, other than a mate," Lilith replied. She had tried not to hurt Ann in the way she said it, and was reasonably confident that Ann knew her well enough not to take offense. Still, there was a bit of sting in the comment, if only because being without a serious relationship had for so long been a mild but nagging omission in Ann's life. Funny how, despite her countercultural views in the politically correct university setting, it was the absence of a significant other more than anything that had made Ann feel out of place at times.

226

"I know how impractical it is. I've got my work. That's my baby, I guess. Still, the prospect of children for me, and most women, I suspect, is in our DNA. I suppress it most of the time. But you said you wanted to talk."

Turning from Ann, Lilith looked up and placed her hand on the back of André's, who was setting their appetizers on the table. "Would you be ever so kind, André, to bring us another round, please?" She looked over to Ann to be sure she was okay with the request for two more drinks before they'd finished their first.

"Certainly, miss," he said, now holding her hand in his. "For you, anything." He kissed her hand before turning away.

When he left, Lilith faked girlish embarrassment. She and Ann shared a laugh at Lilith's simulated infatuation, just as a jazz guitarist began to play quietly. The two took up their conversation again.

"So, Ann, how is your research progressing?" Lilith asked, hoping to make small talk until Eva came on stage. Ann's work was interesting to Lilith, as it was to everyone who had at least some familiarity with it, but now wasn't the time to get into it.

The scope and depth of Ann's work, or more accurately the university's work, was breathtaking. Historically most university labs, regardless of the academic department, were as varied in their research as the professors and the doctoral candidates who worked in them. In Ann's case, inspired by the relatively rapid success through global collaboration of government agencies and universities participating in the Human Genome Project nearly two decades prior, she had persuaded the University of Minnesota to take the lead in organizing a multi-national public/private consortium. Its purpose was to understand the biological mechanisms of brain development and function.

Minnesota's portion of the larger project initially involved a dedicated, interdisciplinary team of biometricians, geneticists, behavioral psychologists, and computer scientists in conducting a new, more massive Genome Wide Association Study (GWAS). This statistical method for determining the relationship between genotypic variance and phenotypic variance helped define the particular gene alterations that account for variation in some manifest characteristic of living organisms. Ann's GWAS study was primarily to identify the most comprehensive list possible of genes that influenced intelligence. Identifying genes that contribute to brain development and functioning was a critical first step, but by no means the only step. Several years back, Ann had begun to convince many of her peers in academia, particularly those at MIT, Harvard, and King's College in London, among others, that Minnesota was the perfect institution to lead this particular GWAS effort, given its history in cognitive psychology and twin studies, and its expertise in behavioral genetics.

Ann had been one of the early advocates for institutional collaboration, given the direction of her own academic career on brain structure and function. So engaging was her personality that she was able to draw in her initially reluctant peers. In time, several large corporate benefactors, coupled with certain government interests, saw the merit of her approach, and funded the effort beyond her expectations. The Broad Institute, a partnership between Harvard and MIT supporting genomics and medical research, was an enthusiastic early participant in the effort. Building on the list of genes to be identified in Ann's study, the Broad Institute would evaluate the effect of gene expression on brain structure development using advanced scanning tools.

Ann's job had evolved in the last few years. She began to lean more heavily on docs and post docs to carry on with the GWAS, while she shifted her attention to coordinating the collective activities of the consortium. By unanimous agreement, she assumed the role of its de facto head, and concentrated her time on building the capability to share the group's rapidly advancing knowledge. Advancements in computer science, particularly in the domain of artificial intelligence and data mining, became as much a part of the group's work as investigations into strictly biological phenomena. Managing information flow and analysis became more than an adjunct to their collaborations, it was the central nervous system.

"You remember me talking before about Genome Wide Association Studies, GWAS, right?" Ann asked, beginning to lean into the subject.

"I do, at least I think I do," Lilith answered. She immediately recognized she was in for a lecture, and regretted broaching the subject. Ann never could resist.

"GWAS associates DNA for a large number of individuals with certain observable characteristics. In our case, we were hunting for the genes that affect brain development and function, with particular emphasis on general cognitive ability — intelligence. Meta-analyses are larger studies through which we amalgamate previously published smaller studies. About ten years ago, we identified just over one hundred protein-coding gene regions that associate with intelligence. We can say, therefore, that the trait for general cognitive ability is polygenic. There are many genes that influence intelligence, though each may have only a small effect … maybe less than 1%. That's not so surprising given how many different types of brain cells and structures there are, and the biological processes necessary just to maintain each cell. Through those smaller studies, we found that the first few dozen genes only accounted for something like less than five percent of the variance in intelligence. So we had a lot more variation in IQ left to explain. We knew from twins studies that IQ heritability ranges anywhere from forty-five to eighty-five percent in children and adults respectively, depending on the study. So, from twin studies, you can see that it's a largely inherited trait. Am I losing you?" she asked, sensing that, once again, she had gone too far.

"I'll let you know, but it sounds like you're about to reveal something juicy," Lilith replied.

"I am. Well, the problem with most of this work is that those GWAS studies didn't go far enough. The sample sizes were too small. We simply did not have enough data. That is, until about five years ago. Part of the problem was the cost and the time it took to sequence the genome. The other problem was getting intelligence data on those whose genome was already sequenced. You still with me?"

Lilith nodded between bites of guacamole.

"The National Institutes of Health a few years ago began a longitudinal study that would ultimately include sequencing genomes for more than one million people. As part of that study, other behavioral traits and health information was collected, including SAT scores, which are reasonably well correlated with IQ. Added to that database is one that is part of the Million Veteran Program, begun in 2011. Though focused on tracking health, genetic information along with results from the Armed Forces Qualification Test were included. Meanwhile, a number of other countries have compiled data from their own studies. Including data from outside the U.S., the total number of the aggregated population in those studies now numbers in the millions," Ann said. "Each time we added more data, we identified more genes that influenced intelligence, and accounted for more of the variation in intelligence." Lilith saw that Ann's eyes were opened wide now.

"And ..." Lilith said, still waiting for the punch line.

"And we were able to complete a meta-analysis of all of these data. Right here at the good ol' U of M! Turns out that we're up to more than twelve hundred genes, and those genes account for somewhat more than twenty-five percent of the variance in IQs. We'd broken through what was once thought to be an intractable barrier. Of course, to this point, the evidence is all statistical, and our work is just the first step in understanding how our genes effect brain development and function."

"Amazing," Lilith said, trying to sound enthusiastic, although her manner suggested that the study was all so esoteric that its importance was obscured. "Ann, I've got to say, your dedication is impressive. But in the overall scheme of things, just what difference does it make? You and I are still single. Though cute, André seems ... well, simple, and the world is still as screwed up as ever."

"I haven't told you the best of it, Lilith," Ann added, now being a bit coy.

"And just what might that be?" she asked, sounding skeptical.

"You know we have the largest interdisciplinary team in the world working on this. It's not just a bunch of statisticians in our lab working on the problem. We were just the folks figuring which genes are in play. That doesn't tell us what each one does, or how they interact with each other, or how they work in individual cells to make the brain function. Look, through our consortium we've got some extremely talented ... I should rather say 'rare geniuses' working on special types of brain imaging technology. Using techniques for gene editing ... CRISPR ... you've heard of that, right? ... we can turn genes on and off, then observe which genes and combinations of genes affect individual brain cell types. We can now actually see those effects under the microscope in living tissue, and in living animals. We can also see changes in development and variation in brain connections, the connectome. As a result, we can also affect the working of neurotransmitters. In short, with these new tools we can actually modify genes, and then visualize the differences in brain structure and chemistry. Today, if you give me a full sequence of your genes, I can with improving accuracy tell you how the structure and function of your brain differs from the average. To top it all off, we are just now at the beginning stages of being able to fairly accurately predict your IQ from your genome. Our hope, in time, is to be able to get within fifteen IQ points … one standard deviation," Ann said, waiting for Lilith's reaction. "At one time, the very best we could hope for was to predict within maybe two standard deviations … about thirty IQ points. But now, we've also learned that it's about more than just identifying individual genes and the proteins they express. We've learned quite a bit about gene regulatory elements, the excitatory and inhibitory elements, and how they affect gene expression. We're also beginning to see how the environment modulates gene expression, what we call epigentics." She leaned back in her chair to see how these claims were received.

"I don't mean to disparage your work. It's just that I don't see what all this has to do with anything. Besides, wouldn't it just be easier and cheaper to administer an IQ test? Learning how the body works is nice, but just what do we do with all this knowledge? How will it improve the lot of man? More importantly, how will it find us a man?" Before Ann could answer, Eva stepped to the microphone and began to sing. Lilith's question floated away on jazzy riff as Ann and Lilith turned their attention to the stage.

After Eva's set, a small trio came on stage. Lilith and Ann settled up their bill and headed to a quiet beer joint near Ann's house. They ordered a couple of beers at the bar, then moved to a booth in the back.

"Eva was fabulous," Ann offered. "I'm so glad you suggested this. You know I don't get out as often as I should."

"I know. That's why I invited you, and yes, Eva was great. I don't mean to change the subject so abruptly, but …," Lilith began.

"But …." Ann interjected. "That's like saying, 'I don't mean to interrupt, but' …. Or, 'to tell you the truth….'

They both laughed, and riffed on contradictory prefaces. Lilith began again. "Now you've got me intrigued by your work, Ann. It all sounds so SciFi, but I suppose it is leading to something important."

"It is," Ann said. "I haven't told you everything. What I'm sharing is pretty confidential. We haven't had our work published yet. As you well know, peer-review is everything. It's the first step, that is, AFTER one of the journals even agrees to consider our research. What I'm about to tell you is very confidential. Do I have your word …?"

Before she finished her question, Lilith gave her resounding assurance.

"We've actually gotten pretty far along, and as I said, we aren't working alone — that is, just in this country and Europe. Perhaps you're aware that we have quite a few Chinese doctoral candidates and post-docs working with us. Over the years, we've developed a relationship with one of the leading Chinese research institutes, the Beijing Genomics Institute. Given the strictures on certain kinds of research in the Western world, we've been able to take advantage of liberties the Chinese offer in their system."

Leaning in, and now speaking in a lower tone, "And what is it that they are doing for you, exactly?" Lilith asked.

"For many years it has been widely accepted in the international community that genetic modifications are impermissible in the germ line," Ann began. Detecting Lilith's confusion, she explained, "I'm talking about editing embryos, or even eggs and sperm. Look, any modification to humans that can be passed on to future generations through normal human breeding is forbidden. Since we still can't always predict with precision what gene editing might do in all cases, it is widely accepted, therefore, that we ought not genetically alter embryos unless those embryos are destroyed within a certain number of days."

"So, you're telling me that the Chinese are doing gene edits that do in fact alter the germ line, and they are not destroying the resulting embryos," Lilith whispered.

"That's partly right," Ann replied. "They'd been doing it for years. As far as I know, they've experimented only on lab animals and domesticated livestock. Some have gotten a bit grotesque in the process. They've developed super-muscled dogs, for example. The deal is, they can get closer to the line than we can. Not long ago, they edited twin embryos, but were widely criticized for it and said they'd stop. But I'm hearing rumors they're still doing it, this time out of the public eye."

"And just what is their aim? What are they hoping to accomplish, and more importantly, what have they learned?" Lilith asked.

"Look, let's just say that their work is, as far as we know, still in the very early stages. We are just beginning to get a hint of what they're up to. Of course, our international partner is not completely under our control. Consequently, they do not share everything with us ... at least in a timely fashion. As luck would have it, I'm going to China next July for a major international conference. I'm presenting a paper on our work and hope to learn more about what they have been doing. Until then, I guess I just can't offer any more information than what I've already shared. Again, promise me you won't repeat anything I've told you." Ann sounded almost pleading now.

Lilith acknowledged that she understood the seriousness of the matter and pledged her silence.

"Lilith, I don't mean to sound like I'm up to some kind of nefarious science. All our consortium hopes to learn is how genes affect brain development. If in the process we could somehow find a way to eliminate autism, low cognitive ability, dementia, or mental illness, for example, it's certainly in everyone's interest. It's just that, you know, the press and the public get kind of crazy when we talk about gene editing. Somehow they think we'll create monsters, or designer babies, or unleash some kind of unknown disease on humanity. Trust me, it's not like that at all."

"Designer babies. Wouldn't mind having one of those myself. Maybe something in beige — with blue eyes, of course," Lilith joked. "At least I once thought I did. I guess I'm getting a little old and too settled for that now. A baby would cramp my style. How about you? You mentioned you might want a child some day," Lilith asked, trying to move the conversation in a lighter direction.

"Oh, geez, I don't know. I'm too married to my work. It's true I once flirted with the idea of adopting — even considered a surrogate, or *in vitro*," Ann confessed.

"You serious?" Lilith asked, incredulously.

"I am, or should say, I was … once. Like I said, I'm too busy now, but what woman hasn't thought of raising a child?"

Ann reflected occasionally on her earlier divorce, sometimes only momentarily, sometimes in a period of prolonged brooding. She'd been thinking about it more lately. She'd seen it as her failure, and the hurt never really healed. While her life til now had hastened on, and her feelings of relational failure faded, more often of late she reflected on what her life might have been. Perhaps it was because her teaching had become more of a pain. Then, too, she could see that her research was coming to an end without the breakthrough she'd been aiming for. Maybe it was her age, perhaps her midlife crisis. She was starting to feel trapped, and was increasingly thinking about how she might make some kind of change. It would be a life freshening. It could mean a shift in career direction, maybe even a whole new career. That would be the ticket. And in the middle of their conversation, even now in this small neighborhood bar, out of place and out of time, these thought once again crossed Ann's mind.

Liz responded, "Well, since we're being completely honest with each other — which I have to say makes me very uncomfortable — I was pregnant once. I'm sure I must have told you this." Lilith said. In spite of her offhand delivery, her eyes revealed something that Ann had never noticed in her — regret, maybe even shame.

Trying not to sound shocked, knowing that might prevent Lilith from revealing some hidden truth, Ann said. "Oh, no, I didn't realize that. What happened?"

"I was in grad school — up to my ass in my studies. There was a guy. You know. The usual. Anyway, I got pregnant. Never intended to. It wasn't a good time for me — for either of us. So, I did what any girl in the same situation would do. I had it aborted. It took me a long time, maybe too long to make that choice. In denial for a time, I repressed it, but in the end, accepted that it was for the best. Turns out the doctor said it was defective in some way. In any event, I wasn't equipped to deal with it at the time. I'm over it now. I really am."

"Oh, Lilith, I'm so sorry. This must have been such a heavy burden," Ann said, but she didn't believe Lilith was over it.

"Well, it was something to get through, for sure. We did, so here we are. I think stronger for it. Anyway, so … it's not likely I'll be having a baby in my future. That's my story," she said as she offered a sheepish smile to her friend.

"I've got to ask you, if you don't mind," Ann continued. "What about the guy? What happened with him?"

"Typical. He disappeared. Wasn't very supportive. You know," Lilith said, now sounding matter of fact. "It's somethingg I had to do on my own. Like I said, it just wasn't a good time, but the fact that it was defective, well, that made it easier to accept. I suppose even if I'd been a married housewife at the time, I would've had to abort a defective child. It's just too hard in this world. You know what I mean?"

Ann didn't answer. She and Lilith looked into their beers for a moment, and then simultaneously took a sip. Ann still hadn't thought of what to say, but abhorring the silence, which to her conveyed a lack of empathy, she offered, "Well, you're not alone in that. Thousands of women terminate pregnancies, particularly those with congenital problems. In some countries, Down syndrome has all but been eliminated. I certainly wouldn't wish to have to make the tough decisions that they do. It's got to be a personal decision, you know. No one can make that for you."

Lilith surveyed Ann's face to try to gauge whether her expressions of support were heartfelt or just throwaway lines. "I suppose that one day your work might find a cure, or a way to eliminate those problems. It'd sure be nice. Women wouldn't have to make such awful choices. As for the men, any way you could reengineer them so they have a bit more sensitivity?"

Their laughter resonated in the near-empty beer parlor. The few remaining patrons stared with glazed eyes at their booth, then turned back to look into their schooners. Ann and Lilith finished their beers, and then Lilith took Ann home.

The next morning Ann rose early and went directly to the lab. Most of the research team would be at work, as would their collaborators in other labs around the world. The group doing gene editing would be busy mixing amino acid compounds. The scanning group would be preparing specimens, while others would be feeding lab animals. Ann's own team would be poring over the data from the last run, assuring that any anomalies could be explained.

Of necessity, Ann played down the explicit goal of her research, which was to develop a polygenic scoring system for predicting IQs from genes. The work was so controversial that its continuation had often been threatened. Considerations of race and class, not to mention the collective memory of the eugenics movement, could be more formidable barriers than the science itself. At times several interest groups tried to stop her research, though most did not really know her ultimate aims. The threat of foreign competition, in particular the Chinese, had ultimately persuaded financial backers to continue their support in the face of such resistance. The only limitation imposed on Ann and her collaborators was that they must keep their work quiet until they were able to demonstrate not just the genetic, but also the microbiological mechanisms of general cognitive ability. When required, the public face of their research had to be on eliminating human suffering rather than suggesting ways to predict and improve native intelligence. Discussion of intelligence too easily devolved into debating systemic injustices, and to an examination of social policy more than scientific discovery. Ann and her colleagues learned to just avoid the debate. As Ann said, "We are conducting genetic research on factors influencing a healthy brain. Our work is foundational in nature, though we hope it may one day lead to improvements in mental health."

In reality, those working with her knew that the real target of the work was understanding the mechanisms of 'g', the general construct for intelligence. It was widely accepted, though not universally so, that higher intelligence confers significant social advantage, at least when considering large samples. Demonstrably, it is associated with greater educational attainment, higher earnings, better health, greater social integration, and in general, better life outcomes. To those who would argue that structural inequities were important in the determination of life outcomes, the scientific community would generally concur, yet they likewise knew that for any given socio-economic class, those with higher IQs tended to have better life outcomes.

◆

Ann's custom was to take the stairs. Just as she turned the corner at the top of the stairwell, she spotted the backlit silhouette of someone just outside her office door. As she approached, she heard snuffling and intermittent sobbing. It was Mark, the cognitively impaired janitor who worked in her building. He sat on the floor with his head in his hands, elbows resting on his knees.

"What's going on, and why on earth are you here on a Saturday?" she asked, not knowing at first how to proceed, or offer comfort.

He looked up at her, tried to compose himself, and then sobbed more loudly. Ann squatted by him and touched his shoulder. "Tell me, what's the matter? Maybe I can help," she said.

"It's my grandmother," he cried. Too distraught to continue, he buried his face in his hands.

She patted him on the back, stood and said, "Why don't we go inside and talk about it," She put her hand on his shoulder. Mark got to his feet, and she guided him to the chair next to her desk.

CHAPTER III

Put one hundred fairly typical people in the same room, and, absent celebrity, it is doubtful that anyone would particularly stand out. Natural variation would be familiar and for the most part limited. Body size and shape would vary somewhat, so too facial features, but few if any physical characteristics would be outside an expected range. For normal persons, there likely would be nothing in their physical form that reveals the extraordinary — nothing that would attract much attention. The same observation could be made if you measured behavioral characteristics.

Run the experiment again. This time put Mark in the room. In mere seconds, he would be seen as markedly different. Only social norms would prevent one's gaze from lingering. Distinctive in stature, build, facial construction, even down to the length and width of his digits, the observable differences would be striking. Interact with him, and one would easily recognize behavioral differences as well. Mark would be an outlier.

At the genetic level everyone in the room would be remarkably similar, unsurprisingly, because the genetic differences among them amounts to something like only .1%. Said differently, all share the same genes, and only a relatively few of their bases would vary from the norm. Given the extraordinarily large number of base pairs in DNA, the degree of genetic concordance for the group would be astonishing

— but for one notable exception. Mark was born with an extra chromosome 21, a third copy, which he inherited from either his mother or father. He has a genetic redundancy, an extra genetic load, and that principally makes all the observable difference.

Chromosome 21, the smallest of the 23 different pairs of chromosomes, is estimated to have only 250 genes out of a total of something like 20-25,000 genes in all of the chromosomes combined. Chromosome 21's 48 million base pairs represent a mere 1.5% of the 3.2 billion in Mark's total DNA, so the overall influence of this single extra copy of chromosome 21 would seem to be correspondingly small. Yet this relatively small difference, given the extra dose of proteins expressed by this third set of genes, produces outsized effects. Complications associated with that over-expression can range from mild to severe, and can manifest in the following conditions: thyroid disease, heart defects, infectious disease, sleep apnea, obesity, seizures, hearing loss, premature aging, skeletal problems, leukemia, visual problems, gastrointestinal abnormalities, early onset Alzheimer's disease, and perhaps most notably and always the case, severe to moderate cognitive impairment. Fortunately for people like Mark, there are medical interventions that can ameliorate many of these conditions, -- that is, with the exception of impaired cognitive ability.

Mark shared distinctive physical characteristics with others having Down syndrome. He had flattened facial features, a smaller-than-average head size, a short neck, small mouth with an often protruding tongue, a small nose, upward-slanted eyes, small ears, a single crease across the palm of his hands, a deep groove between his first and second toes, and poor muscle tone at birth.

Yet, in every respect, Mark and persons like him are fully human, indistinguishable in general form and function from that which makes one human. He eats, he talks, he ambulates, takes care of himself, learns and exhibits the same range of emotions as everyone else. He has ambition, hopes, dreams, sexual desires, frustrations, private thoughts, and even secrets. And for reasons that have yet to be teased out from questions of nature vs. nurture, he and persons like him often exhibit a greater capacity for joy and a desire to love and to be loved than most other humans.

Lest it be thought that though rare, genetic anomalies such as his are unique, every human has genetic variation and misprints. In fact, there are something on the order of 10 million SNPs (Single Nucleotide Polymorphisms), common variants from the norm in human DNA. That's one in every 300 bases (letters A, T, C, G) on average that are 'misprints.' They are common enough, and aside from providing for much of desirable human variation, they sometimes cause adverse consequences. But it's these SNPs that combine to make everyone unique. In sum, human biology is designed to produce variation.

◆

Mark, who was thirty-five years old, was responsible for cleaning the top two floors of Ann's building. That's where professors had their private offices and graduate assistants shared separate office space. Mark cleaned white boards, unless of course there was a note: DO NOT ERASE, which he could read; emptied wastebaskets; vacuumed carpets; and mopped the hallway floors. He knew the routine well, could be counted on to show up for work on time, and could satisfactorily complete his tasks, albeit somewhat slowly. After training in janitorial work by a local sheltered workshop, he was placed at the university some ten years earlier. During his first six months, Mark's work was closely supervised. Upon satisfactory completion of an extended probationary period, he was trusted to work independently. Accordingly, he was paid like all the other college janitors with the same level of seniority.

His limitations aside, Mark's loyalty, reliability, and affability were of consequential benefit in an occupation that suffers from frequent turnover. He had an outgoing and friendly demeanor and was willing to take on new challenges. Owing to these positive traits, his presence at the university was not only accepted, it was welcomed, and he thus became an integral part of the team.

His unlikely presence at a well-known research university was readily accepted, not as a gratuitous gesture of inclusion; it is more accurate to state that he was appreciated for the uniqueness and value of his personhood. Though rare, occasions of antisocial behavior were directed at Mark. Obnoxiousness can be found most places. One such event happened when a new young janitor was assigned to work in Mark's building. Bristling with assumed authority, particularly on having been forewarned that his coworker was cognitively impaired, this young punk was observed trying to take advantage of Mark's good nature. Likely, his bravado was based on the assumption that no one would witness his misdeeds. Fortunately, Ann happened to see the little drama play out.

"Mark, you need to mop my floor today. You missed it yesterday. I have to go to lunch early again and can't do it. I don't want to have to tell you again!" the miscreant barked.

"That's not my floor. You know it's not my floor, Kent!" Mark protested. Mark could be quite defiant when pushed, in his case never in the more cowardly, passively aggressive sort of way.

"I'm not going to tell you again," the bully repeated. "If I have to report you to Ralph, you're going to be in big trouble."

"Look, K..K..Kent," Mark said, "if you think I'm going to c..c.. clean your fl..fl..floor, you can b … b … bite me." Mark turned and walked into the janitor's closet. It wasn't clear to Ann whether he intended to hide there to avoid further conflict, but she was quite certain that he was capable of a more apt response when he emerged with a mop handle and advanced toward his tormentor. At that, the bully turned and walked away in haste. He could be heard muttering to himself as he bounded down the steps and out the front door. Maybe the bully had seen Ann's head peering from around her office door; or, then again, maybe it was Mark's assertiveness. Either way, the bully did not return, and Ann decided right then that Mark was special.

◆

"Why don't you sit here in my chair, Mark, and tell me what's going on," Ann directed.

Between gasps for air, he told Ann that that he thought his grandmother might be dead. He had found her in bed that morning, and she would not wake up. Not knowing what to do, he ran in to work, hoping to find someone familiar he could tell. Fortunately, Ann had come in early and was able to intervene. Together they rushed to Mark's grandmother's apartment, located above a nearby convenience store. Ann confirmed the grandmother's death and called 911. During the wait and the inevitable questions that followed, Ann stayed with Mark. When the ordeal was over, Mark gathered a few of his things, and Ann took him to her apartment, where he would spend the night. It had taken that evening and the better part of Sunday to figure out the name of Mark's social worker, whom Ann would call on Monday. She and Mark agreed that it would be best for him to return to work on Monday.

As it turned out, an immediate out-of-home placement was not easily arranged, given the lack of a group home opening near Mark's work. The only facilities that had room were in suburban locations too far from his job for him to make a timely commute. Foster home placement was deemed unlikely for the time being. Though the plan was irregular, Mark's social worker agreed to let him stay with Ann, as Mark had no family in the area.

Over the next week, Ann was able to keep her normal class schedule and office hours, and managed to check in on her lab with some regularity, at least so that her work could carry on. Mark proved to be quite self-sufficient in her absence and could be counted on to prepare basic meals for himself and help keep her apartment clean. In fact, his presence made it possible to catch up on housekeeping chores that had been neglected.

When she got home later in the evening, Mark was well fed, at least with microwavable meals, and would usually be found watching a favorite TV show or playing checkers against himself. Mark liked to say he always managed to win. Ann would slip into an oversized college sweatshirt and loose-fitting basketball shorts, then flop on the couch with a glass of wine. She and Mark would talk over their day or play board games. Their conversations were light, uncomplicated, and always, always elevated her mood.

After a week of this new routine, Ann thought it appropriate to begin prepping Mark for the inevitability of his moving to a more suitable living arrangement. That really meant a group home, but she didn't want to alarm him with the idea so directly, because it was obvious that he was now quite comfortable living with her. He finally did agree that such an arrangement might offer benefits, inasmuch as he could be living with people closer to his own age and who shared his interests.

Ann briefly flirted with the idea of adopting him, or becoming his legal guardian. It could have been the freshening she was looking for, but would have complications. She wasn't sure how the legal process worked, or even if it could work, but the idea had, if only for a moment, some appeal. In the end, she rejected the notion because of the demands of her job. She rationalized that she might not be able to provide for him once she retired, and would need all of her resources just to provide for herself.

She thought it was funny how living with this thirty-five-year-old, cognitively impaired man seemed so natural. She expected that though his IQ must have only been in the forties, he was in nearly every respect capable of carrying on an adult human life. He could care for himself, work, have and share feelings, and add value to the life of another. It didn't seem to matter, and in fact it was welcome, not to have deep intellectually, morally, or emotionally charged conversations. Yes, there were occasional disagreements, though they were minor and fleeting. The most important thing was that they were able to relate to each other on a very basic level. Besides, she got enough of the other kinds of interaction at work, and too often the nature of those interactions was coarse, aggravating, or both. For a moment, she reflected that IQ was overrated. Weren't most of the big problems in the world, the really big problems, caused by people with high IQs, and especially those who also had an elevated sense of their own capability? Some people liked to blame religion for the world's troubles, but most big problems were caused by well-educated people who were convinced they had the answers, but who were simply wrong. There was something to be said, she thought, about just being with a simple, kind, and happy person.

Mark's time with Ann settled into an easy rhythm, as each adjusted to the other's routine and as they got to know each other more fully. By the end of the second week, Ann began to notice Mark's small idiosyncrasies, some of which could be grating. There was the matter of his mouth-breathing, which even in her most accommodating mind-set she found annoying. Then, too, there was the manner in which he ate. Chewing with his mouth open and joyously savoring his food by making loud smacking noises was something that, even with mild correction, she was not able to modify. It was good that she could be away from him for the better part of the day, so that when they were together in the evening she could still enjoy his companionship. In the end, Ann accepted that living alone for so long had made her a bit less tolerant of others.

Ann contacted Mark's social worker multiple times over those first two weeks, primarily to talk about options for his future. While the idea of adoption and guardianship were briefly discussed, it was apparent that such an arrangement would be impractical, so the worker continued to seek an alternate living arrangement. Midway through the third week, both she and Ann came to the conclusion that it might be better to find any group home that could take him. They'd just have to help him locate alternate employment. They knew that the longer he stayed with Ann, the harder it would be for him to separate. Either way, finding suitable employment would be easier than finding a group home near the university.

His grandmother's funeral and burial arrangements were finally made with the social worker's assistance. Mark managed to get through both with what to Ann was surprising equanimity. Ann learned from Mark's social worker that his acceptance of what had momentarily been a totally upending event was nothing out of the norm for individuals with Down syndrome. Emotions, though at times amplified, were fleeting in most cases. For people like Mark, the only sentiment that lingered for longer intervals was a grudge, and even for that, the cause was soon enough forgotten. Ann and the worker managed to have a laugh at the idea that they, too, should be so blessed.

When at last the time came for Mark to move, an eventuality that Ann and the worker had prepared Mark for, the move went off without a hitch. The home into which he moved was all the way across the metro area, so Ann would not see him as often. Ann did call and send post cards, which he especially loved, and took him to dinner on special occasions.

Mark adapted well to his new home, particularly given that there were other young men his age living there. A janitorial job was found for him at a vocational-technical school just a short bus ride away. Later, Mark met a girl at a dance, and before long claimed her as his girlfriend.

Ann was customarily self-effacing about taking Mark in, though to her friends she sometimes dramatized her sense of guilt at initiating his move. Lilith was never one to let such virtue-signaling slide. She once chided Ann in a way only she could. She said that Ann's false modesty and manifest anguish looked to others more like preening sanctimony than guilt. It was a harsh rebuke, but balanced with occasional praise eventually restored in Ann a more balanced assessment of the whole affair. Her act wasn't noble; it was just the right thing to do.

During the first week after Mark moved out, Ann was called into her supervisor's office for what was to be the most unsettling meeting of her tenure with the university. Over the last month, unbeknownst to her, an investigation of Ann's classroom conduct had been carried out by one of the university's many standing faculty review committees. Their work had been accomplished with surprising secrecy, at least for a university, where gossip was the *lingua franca*. The student who had walked out of Ann's class in a fit of pique over alleged racism and several other 'isms' had indeed gone to the dean to complain. So heinous were the charges considered that the faculty review committee immediately began conducting student interviews. They managed to obtain cell phone videos of the incident in question. The portion of Ann's lecture in which she asserted that there was a genetic basis for racial categories, and the part in which a student commented on the fallacies of climate science, were included in the video. Needless to say, these two were most troubling to the administration. It was inevitable that Ann would be called on the carpet.

"Ann, how are you?" Dr. Kerner began. "Why don't we sit over here?" he suggested, gesturing to the low-slung couch and adjacent high-backed side chair. These were situated next to a large coffee table on which were displayed several expensive Chinese jade and ivory figurines. Ann took a seat on the couch near the side chair. "I trust your research is going well," he continued, stalling for a time to gauge her mood. "And your classes," he continued, "I trust everything is fine with your schedule this term."

"Yes, thank you for asking. Everything is going well," she said. She kept her responses short so that they might more quickly arrive at the point of the meeting. One-on-one meetings between this department chair and his associates, they both knew, were not something that anyone in her field had reason to expect. There was an exceedingly long pause after her response, during which time the department chair leaned forward and carefully rearranged the coffee table figurines nearest him. Once done, he assumed his usual imperious posture, with palms touching, fingertips touching the area between the tip of his nose and his upper lip. He tapped several times, and then began.

"You may wonder why I've asked to meet with you today," he said, then added in a faint attempt at humor, "it not being anywhere near Christmas." She did not smile. "It seems we have a matter."

"A matter?" she asked.

"Yes, quite so," he responded. "Let me not dally. Although over the course of your tenure here, your teaching has always garnered high marks — I must add that I, too, have seen your teaching on more than one occasion, and concur with those assessments — it seems of late something altogether different might have altered that view."

"Oh, and what might that be?" she said, growing impatient with his tiptoeing.

"We have, I should rather say, the university has had a number of complaints about your interactions with students in the classroom. Specifically, there have been allegations that you have taken positions that," he paused, "… that run counter to the values of this university. And because of those allegations, we have of necessity undertaken an investigation."

"Excuse me, Dr. Kerner, but can you be specific about those allegations?" Ann said. Her upper lip tightened as she spoke.

"I can," he replied. "Specifically, it has been reported, and confirmed by more than one student, that you have espoused a view that there is a biological basis for race. Moreover, it is alleged … er … that you have seemingly advocated positions in support of eugenics. This, I think you would agree, is not only counter to our values as a university, but morally repugnant," he concluded.

Ann said nothing.

"As I say, we have conducted a thorough investigation, an investigation which also included a review of a video tape of the incident in question. Further, we have reviewed a syllabus for your course, spoken as I say to a number of students, and reviewed student notes from your lectures. While we can't reasonably conclude that the student's assertions are altogether factual, we can say that you have approached a line, one that we do not wish, indeed, shall not be crossed. Do I make myself clear in this matter? Do you have anything to say for yourself?"

"I take your point. I can assure you that I have neither advocated for eugenics, nor done anything other than provide an accurate history of the origins of the field as a preface to my course on behavioral genetics. As to race, my assertions were factually correct, and are supported by many leading scientists in my field. While those assertions may be unwelcome to the often overly sensitive ears of ... of ... some of our students, they are not claims from which I intend to withdraw. If we are to speak the truth, I, and those like me on this faculty, shall not be intimidated. Do I make myself clear?" Ann, by this time was quite puffed up, even surprising herself at her own temerity.

"Yes, Ann, I think we do understand each other. But let me just caution you again. As you approach your subject matter, it is in your interest to be careful, or at a minimum to prepare your students to more fully understand what positions you are taking, and why. You have been an excellent member of our faculty, and I wish only for you to continue your good work. Now, if you will excuse me, I have another engagement across campus," Dr. Kerner concluded. He rose with some effort, and showed Ann to the door.

The time it would take Ann to walk back to her office was insufficient to calm herself, so she took a seat on a bench shaded by a tree that must have been there since before the college was founded. She needed time to process what it was that she had just been through, and as was her way, to figure out what it meant for her actions going forward. She realized that the encounter could not be easily forgotten. Given the indirect manner in which the university customarily meted out discipline, she must in some way take heed. She knew that what she had said in class, though true, was not managed in a way in which she usually conducted her class. Things had gotten out of control. Students were different today. They wore their sensitivities on their sleeves, but more ominously, they acted on them. What was still true was that students, even though they were more generally aware of the world around them, were no more mature or wise than students from generations ago. Devoid of knowledge or life experience that might otherwise have informed their actions, they were more likely to act on feelings and abundant misinformation that lay only a few keystrokes away. While pondering the higher education types that might have made up the faculty review committee, those thinking men and women of letters, bile of contempt arose in her throat. Till then, she'd always managed to suppress feelings of repulsion, though it had long been apparent that, like her students, many of her colleagues were too easily guided by feelings of moral superiority and rectitude, and adhered to a code of political correctness that now infected nearly every institution in the country. Calming herself, she reflected on the changes she had noticed in herself — the occasional crankiness, her abrupt response to students, and the yearning, which seemed to come out of nowhere for a dramatic change in her life.

Needing to vent, she called her brother that evening. Though irregular communicators, they exchanged calls with sufficient frequency that they generally knew what was going on in each other's life. They were as close as siblings could be, and this not least because they came from the same gene pool. But, they were also as different as two people could be in outlook and temperament. Andrew, a practicing clinical psychologist, was easy-going and unflappable, even to the point of teasingly being called phlegmatic by his sister. Andrew wasn't roused by much. Clinical psychology suited him, and his work with veterans suffering PTSD was a calling. Probing the frightening and sometimes heinously ugly thoughts of those suffering from guilt and fear-based trauma required of him an ability to separate fantasy from reality, and to hold onto a view of mankind that was uncommonly positive in an age of politically charged news. The torment of his clients in no way manifestly affected Andrew, as he never ceased to search for that which was positive in the other. It was belief in the goodness of others, more than any exceptional ability as a psychologist, that helped his clients. It was his emotional generosity that managed to worm its way into, and alter the thought processes of those he was paid to help. The fact was, he would have done his work for free, especially given that his wife was a well-paid partner in a national law firm.

"Hey!" Ann began. "What's up?"

"Oh, hey, Sis," Andrew answered. "Just watching a ball game. What's new with you?"

"The usual, you know," she said. Ann usually didn't approach a subject cautiously, like most people. She considered that a long ramp-up to saying what was on her mind was a waste of time. "Got called on the carpet today … by the chair of my department. Never in fifteen years had that happen."

Andrew observed, "That must have been a shock." It was reflexive for him to parrot the emotions of a client in this way. Acknowledging the feelings of those with whom he conversed sometimes got to the point of being annoying, especially with those whom he knew well.

"Cut the therapy crap, Andrew," she said now, sounding peeved. "Of course it was a shock." She knew, and he knew, however, that his response to her opening statement would elicit more information. "Turns out one of my students called me a racist."

"Projection, maybe," Andrew replied. "Or else a sign of limited argumentation skills. It's in the news every day now. Seldom much to it. Mostly white noise. Oops, I can't say that. Social justice warriors are hiding behind every lamppost. I wouldn't worry about it. How did the meeting end?"

"Just like every faculty meeting — no resolution. I'm just supposed to watch myself," she finished. "Anyway …," she left the word hanging as a signal to change the subject, her short vent having fulfilled its purpose.

Persisting, Andrew observed, "Nothing to be too concerned about. You and I both know that in the citadel of academic freedom …."

Interrupting, she said, "Citadel of academic freedom … that's like saying Mayan Riviera."

Getting the simile, Andrew said, "Oh, like after the last cartel mass murder there. Or New Jersey, the Garden State."

"Okay, okay, enough. The irony is that in a place of so much so-called academic freedom, so much conformity of thought is expected, at least on certain topics," Ann harrumphed. "It's not the faculty or administration I fear. The most they've ever done is orally reprimand, and even that was for an adjunct. The thing that concerns me most is a mob of ANTIFA thugs wearing black hoodies disrupting my classes, or worse yet my lab. You remember what they did at Berkley to Charles Murray when he wanted to give a talk on his book, *Coming Apart*. They hate him for *The Bell Curve*."

Trying to interject, Andrew said quietly, "Ann …."

Not hearing, she went on, "I just can't afford it, and can't rely on the administration to stop it. They're more ready to curtail my speaking uncomfortable truths than to put a stop to thugs who would try to silence anything that doesn't agree with their nihilistic views."

"Ann …."

"What, Andrew?" Her tone was harsh, and impatient.

"You're ranting," he said, almost whispering. Shifting gears abruptly, he asked "Say, did I tell you that I heard from that genealogist we hired? Apparently, she thinks she found our great-grandfather on our grandmother's side. She's not sure yet, but she thinks he ended up somewhere in the Midwest, at a state hospital. I should know more in a couple of weeks. She said she needed some time to confirm her theory, and said that unlike most people who do genealogy, who jump to conclusions without documentary evidence, she would have to do a bit more digging."

"Oh, swell. We've got mental illness in the family. Explains a lot. Anyway, you'll let me know as soon as you find out anything?" Ann responded, now more composed.

"I don't think he was institutionalized, but I guess we'll find out, and yes, of course I'll let you know what I hear," Andrew replied. "So what else is going on? How's your research work coming?"

"I shouldn't tell you this just yet, but we've hit on something rather significant," she said, now sounding more upbeat.

"Oh, and what is it that you will tell me that you shouldn't?"

"You remember I told you a while back that we're on our way to confirming some more of the genes affecting intelligence? We've gone beyond that to develop a new polygenic model for predicting intelligence from those genes. Turns out to be quite an improvement."

"Uh huh," was all he said, sounding a bit indifferent to the idea.

"Come on, Andrew. This is important," she said peevishly.

"Yes, and …," he said, again relying on his therapeutic approach to conversation.

"Look, if we understand the biological mechanisms for intelligence, we ought to be able to make useful predictions from what we can observe in the genome. It's not about eugenics. It's that we could understand how our genes largely make us who we are as individuals. It'll also help settle some arguments about nature vs. nurture."

Andrew was silent. Ann said nothing. At last, Andrew spoke up, "Okay, but then, how will we use this?"

"I'm surprised at you, Andrew. Forgive me. That sounded insulting. I don't mean that, but look, if we understand these genetic mechanisms, we might begin to develop the means for eliminating all kinds of suffering. You know as well as anyone that there are many inherited brain disorders. If we had the means, and I think that we will someday, to address these genetic anomalies, we might begin to eliminate suffering from any number of other congenital brain disorders. I ask you, what would be the problem with that?"

"Ann, I don't mean to sound pessimistic … I think you know that I am not, but you know as well as I do that if we could use these techniques for advantage, I mean real human advantage, or even dominance, someone will," Andrew answered. His reply sounded a bit contentious, especially for Andrew, but he wasn't finished, having hit his own rather high threshold for intolerance. "Ann, let me just ask you this: If you had a disabled child, say one with some cognitive disability, and say you loved that child deeply, as I'm sure you would, how could you possibly think in any way of altering that child through genetic engineering? Or worse yet, terminating the pregnancy of a fetus you knew would be born with some disability? Do you think lower intelligence needs to mean suffering?"

"Andrew, I think you know me well enough to know how I'd answer that. Anyway, we are nowhere near being able to do what you suggest. All we're trying to do right now is untangle the mystery of how our genes partly make us who we are. There's nothing more to it than that. In any event, there are widely agreed-to international standards on just what research we are able to do, and how we are able to do it. We just have to rely on the willingness and integrity of those in the research community to comply. So far, we have nothing in particular to be concerned about. Oh, and as for a baby, as you probably can guess, I'm not exactly in a position to even think about that."

"Come on, Sis. You and I both know you've flirted with the idea from time to time. You've told me so yourself," he joshed.

"At my age, and with my workload, it's an impossibility. I've come to accept it. That said, the process of making a baby, well, I won't go into it, but …. there is no one at this time," she laughed, trying to move the conversation in a lighter direction. "It's funny," she went on, "not long ago my friend Lilith and I were just talking about this."

Andrew laughed, again reflecting Ann's emotion, then asked, "And what did you and Lilith agree on?"

"We agreed to finish our beers and go home … alone!"

Ann could hear a voice in the background on Andrew's end. She recognized it as Andrew's wife. "Hey, listen, my wife just got home. I've got to go. Good talking with you, I …."

"You go, Andrew. Please let me know what the genealogist finds out," Ann said.

"I will, and hey, don't let those ol' farts in academia, or the little snowflakes, get you down. You're the best professor I know."

CHAPTER IV

Ann was surprised by how much her temporary role as surrogate parent for Mark had taken her away from thinking about her work, but she soon got back into her former routine. She seized the opportunity to offer David a part time research assistantship after they finally got around to their previously scheduled meeting. He would help manage the rapidly accumulating administrative activities that came with her role as nominal head of the consortium. At Ann's inducement, David decided he would apply for a doctoral degree program in behavioral genetics. His background in psychology, statistics, and information technology would be useful, she asserted. Ann promised to exercise her influence if he was serious.

Ann's group had nearly completed the second iteration of their polygenic scoring system, though it was still far from achieving the level of predictive power they believed possible. It was progress, but still far from a satisfying result. While Ann's latest study was underway, the science of genomics had progressed so that it became possible to extract RNA from individual cells. Measuring this transcribed form of DNA gave scientists a way to assess the degree to which specific genes were being expressed within specific kinds of individual cells, which provided a way to evaluate how well those cells develop and function. As the techniques of RNA sequencing were refined, it became possible to quickly analyze gene expression within individual cells on a massive scale. This new approach was beginning to show that it was not enough to know whether highly intelligent people had the requisite genetic sequence. It was also necessary to know how genes functioned at the level of individual cells.

Meanwhile, the Beijing Genomics Institute had begun its own research. They were using a form of the gene editing technique called CRISPR to evaluate the effects on brain cell development and function. Genetic alterations were made in living tissue, both *in vitro* and *in vivo*. Testing modifications in cell cultures and in lab animals gave Chinese scientists a new way to measure the effects on actual brain tissue and, in turn, cognitive ability. Lab animals undergoing such alterations would, of course, be destroyed or not allowed to breed.

Despite the widespread use of CRISPR in research, it was understood that editing genes *in vivo* might result in adverse effects elsewhere in the body. Genetic expression requires a delicate and synchronous balancing act, which was still not well understood. Outsized risks remained in the degree to which gene editing could be safely used. This was particularly true in human research, and especially in the germ line, where pathogenic edits could be passed down to future generations. For this reason, international norms converged around an expectation that human germ line research was to be strictly controlled.

After David started at the lab, Ann took him around to meet some of her research assistants. "David, I'd like to introduce you to Michael Du. Michael, David just joined our group. David is just wrapping up his Masters degree in psychology, statistics and information technology. He'll begin a doctoral program in behavioral genetics this fall, and will be working with us part-time. I need him to help me get on top of the mountain of communications work with our research partners. Michael is our database and artificial intelligence guy ... our AI guru and all-round tech guy. For now, Michael is the person to set you up on our systems. Oh, Michael, if you could stop by my office in an hour or so, there's a couple of things I'd like to go over with you," Ann said.

"Nice to meet you, Michael. I'll be back as soon as …," David started to say, then looked to Dr. Langley.

"As soon as we finish a quick tour and introductions. Shouldn't be more than forty-five minutes," Ann finished.

Expecting to see pipettes and Petri dishes, maybe a few Bunsen burners and some beakers, David was surprised to see only a few devices about the size of laser printers and a very warm back room filled with computer servers. Mostly he saw desks stacked with paper on which were printed blocks of the letters ATCG in seemingly random order. He noticed that some blocks had been highlighted. When they finished the tour, Dr. Langley led him back to her office. "I realize that saying you'll be helping with administrative tasks is vague. Of necessity, it has to be, and we'll sort that out over time. Suffice it to say that I may ask you to do a wide variety of unrelated things as time and circumstance dictate. I trust that is okay."

"That's fine," he said. "I'm eager to help any way I can. For me, everything right now is a learning opportunity. I've got a question, though. I was expecting to see more of what I think of as lab equipment. Mostly what I see are computers and that Illumina machine, whatever that is."

" I think that's what surprises most people. A lot of what goes on in the field today, and especially here, is computer driven. That small box you referred to is a gene sequencer, though we don't do much sequencing work in our lab. Statistical analysis is where our work is. The volume of data is now so large that humans couldn't begin to process it. As you know, we have had to compare first thousands, then hundreds of thousands, and now millions of samples of sequenced DNA against various phenotypes, in this case intelligence, to see if there is any noteworthy association. We're looking not just at coding regions, but also non-coding, and regulatory regions of the genome. It takes complex algorithms and multi-dimensional analysis to do that, along with a staggering amount of computing power. Unlike the old days, scientists don't so much hypothesizing. Data analytics does that for us. Data drives hypothesis formulation. When we think we have a connection, we use advanced imaging, 3D modeling, and genetic engineering to test these hypotheses. Then, by turning suspect genes on and off, we're able to observe the effects.

"And yes, it is probably true that everything will be a learning opportunity for you. Now, to the heart of the matter: what I'd really like you to concentrate on, and Michael will be your partner in this, is to develop a system that enables our partners to access the rapidly accumulating pile of research plans, data and results — even those things that are preliminary. As you might surmise, we've many reports to share among our formal partners, since we are the designated nexus for all such reports. The field is advancing so fast it's hard for anyone to know exactly what is going on. Work is being unnecessarily duplicated. Contradictory findings are not shared. Different terminology is used to describe the same phenomena. No standardized means of information sharing have evolved beyond less than timely peer-reviewed journal articles. What we need is a structured database … a virtual library for the consortium's work. It's simply too important for us not to have some efficient means of mining what we already know." Ann paused to see how well David was absorbing all this, knowing that she had a tendency to completely disgorge her thoughts in a way that could overwhelm her listeners.

"I can see the problem," David replied. "I've been thinking about it myself, just from my own outside reading. I can see all this leading to some kind of structured database with intelligent analytics capability."

"Exactly! David, I knew you were the right person to take this on," she said, smiling broadly as she reached across the corner of her desk and tapped his knee with the palm of her hand. "All of this is just sort of preliminary. Fostering information exchange is just one problem. It's one we have to take care of quickly. More important … much, much more important, I think, we need to think about the use of artificial intelligence, structured learning, deep learning and all that. I'm no expert, but Michael is a good as anyone in the field, though perhaps not so well known … yet. That's why I hired you. We need someone with both subject matter expertise and working knowledge of coding and structured databases to work alongside our AI guru.

"The future of untangling the mystery of brain science and genetics will depend on how we use technology to make connections that we now don't see. The sheer volume of data is beyond our capacity to process, and is accumulating at a rate we can't possibly comprehend. Scrubbing, ranking, formatting, decomposing and scaling data, perhaps even reducing data, all has to be a part of it. Beginning to figure this out has got to be our ultimate goal."

Looking to bring the meeting to a close, Ann said, "I think you are on the right track. Perhaps you can give me some preliminary thoughts, say, within the next couple of weeks, on how we can start to go about this. Obviously, you'll need to work with Michael … in fact, that is what I'll be meeting with him about in the next few minutes. If after getting to know our people better and getting a sense of some of the research that comes our way, you could give me an outline of how we might begin to structure all of this, I'd be very indebted. I'll check in with you in a couple of days to see whether there are questions I can answer. Oh, and one last thing: I don't expect you to punch a clock, if you know what I mean, though you will have to fill out time sheets. Feel free to get in your twenty hours whenever you find the time in your schedule. Sharon can help you with any of the personnel matters we need to have taken care of, and she can get you a security badge, and whatever office supplies you'll be needing." Ann looked at her phone, "Oops; I've got to run to class, and so do you. I'll see you there."

After Ann dashed off, David walked to the back of the lab where his desk was located. He sat for a brief moment to think about what he had gotten himself into. He panicked at the realization that he didn't have the depth of knowledge he would need. Still, Ann seemed to trust him, and for now that would be enough for him to press on.

"As I've previously mentioned," Ann said to the class, "the heritability of intelligence is quite observable by even laypersons. Comparing patterns of similarity, or concordance, as we say, between monozygotic twins, who share 100% of their genes, and dizygotic twins or non-twin siblings who share on average only 50% of their genes, we can quite easily observe the influence of genetics.

"This can best be illustrated with a story, one that also raises serious ethical questions. In 1980, Mark Shafran, then nineteen years old, drove himself to a community college in New York, where he was to begin his first year. On arrival, he was greeted enthusiastically and familiarly by total strangers who insisted on calling him Eddy. One such person was so welcoming that she gave him a kiss as if they were long-lost friends. As Mark settled into his dorm room, a boy appeared in the doorway and likewise greeted him as Eddy. On discovering his mistake, the boy at the door realized that Eddy, his close friend from the prior year, had a doppelgänger. So struck was he by this fellow student's resemblance to his friend Eddy, the boy suggested that he and Mark should drive at once to New Jersey and meet the real Eddy in person. When they arrived at Eddy's house, the two look-alikes instantly recognized that they must be related, being strikingly similar in appearance and manner. When they learned that they shared the same birth date and were each adopted, their suspicions were confirmed. When media stories and pictures of the improbable meeting of these twins hit the press, another woman observed that her adopted son shared an uncanny resemblance to the other two. Ultimately, the three boys met up and discovered they shared a startling number of similarities in looks, tastes, educational achievement, and mannerisms. Perhaps owing to the media circus that ensued, it wasn't long before the triplets' adoptive parents untangled their common history. Separated at birth in 1961 by an agency that specialized in Jewish adoptions, they were part of an experiment to untangle the influences of genes versus environment. Over the course of more than twelve years, they had been part of a study conducted by Dr. Peter Neubauer and Dr. Viola Bernard, in which placement of the triplets in distinctly different socio-economic environments was orchestrated to help settle the question of whether the rearing environment or genetics was more important. Their physical and mental development was tested, measured, and recorded over the ensuing years.

"The boys' adoptive parents, all Jewish, represented three distinct social classes. One father was a doctor, one a teacher, and the last a blue-collar worker. When confronted, the adoption agency failed to explain the real purpose of the forced separation, and offered only that placement of twins and triplets within a single home was nearly impossible. To some in the field of psychology at the time, separation of such children was considered good for individual identity development. Ultimately, the decade-long research project was terminated. The report of its findings was never published. Fearing lawsuits, all files were sealed until 2065, when all subjects would presumably be deceased.

"That there were so many marked similarities between the three boys was hardly a revelation. Anyone having close familiarity with monozygotic twins would readily agree. Yet the degree of similarity was so striking, especially given that they had been reared apart in homes that presumably offered varying degrees of enrichment, the media could not help but be intrigued by this compelling experiment, however outrageous its design.

"Since the beginning of recorded history, including mythology and early literature, through the observations of Galton, the horrors of Mengele's Nazi-sponsored research, and the more recent analysis by many others, twins have been the subject of much study, speculation, and experimentation for what they uniquely offer to science. And from that work, we have at least gotten a glimpse into the complex relationship between nature and nurture, a relationship which today is opening up to us at a stunningly rapid rate." Professor Langley paused, then turned on the projector.

"Now, as you can see on the screen, with respect to IQ, identical twins, as they are commonly though inaccurately named, are more alike, more concordant, than fraternal twins or non-twin siblings. The distinctions are quite apparent, and this pattern also holds up across a surprising range of behavioral traits, although not to the same degree as intelligence. Even allowing for the fact that, in the studies represented by these charts, all of the matched subjects were raised in the same households, one can judge that more than environmental factors are at work. Sharing the same DNA has to be a contributing factor to the higher degree of concordance between monozygotic twins versus other siblings. I'll say more about heritability again, but for now let me just say that it is not solely a measure of genetic determinacy. Neither is it a statement about the percentage of an individual's IQ that is determined by their genes. Rather, it is a population statistic. So we cannot say, for example, in the case of particular monozygotic twins, that 70% of their intelligence comes from their genes. Neither does it say that 30% of their intelligence comes from their environment, or that intelligence is fixed at conception. What it does mean is that, on average, the detectable portion of genetic influence on the variation in intelligence among monozygotic twins at a specific moment, in general is around 70%."

David, who had walked into class late and missed the first part of the lecture, raised his hand. "Question ... are there any data which compare these populations to adoptees? Because they come from genetically unrelated parents, but share the same environment as biological siblings, might any such data show that they are even less concordant than regular siblings, and that the shared environment is not so important?"

He had hardly finished asking the question when Dr. Langley put up a slide with bar charts comparing concordance between mono-zygotic, dizygotic, non-twin biological siblings, and adoptees. It showed a declining degree of concordance in these respective groups. "That's exactly what we find when we compare these groups, David." She didn't comment on the perceptiveness of his insight; it did not require comment. It was obvious to everyone in the class. "One thing I hasten to emphasize ... you can see that even for monozygotic twins, concordance is not 100%. For that reason, we prefer the more scientific terminology to 'identical twins'. In fact, there are no 'identical' twins. And you have to also conclude that there are influences other than genetic mechanisms, as we understand them, that influence pheno-types, even in monozygotic twins. "

"If there are no other questions or observations, let's move on. In 1974, Dr. Thomas Bouchard, now retired from the University of Minnesota, published a paper on his rather remarkable study of twins reared apart. Let's watch a video of an interview with him as he recounts his encounter with one of the subjects of that study."

When she turned up the lights, she looked around the room to see that the students were mesmerized. "Well," she began, "I see that got your attention. What are your observations? Yes, Collin."

"This is almost incomprehensible to me. It's like an episode from that old show 'The X Files'. How is it possible for twin brothers, raised in such drastically different socio-economic circumstances, to both chew their fingernails, drive the same make and model of car, and have the same hobbies? They didn't even meet until just before Dr. Bouchard brought them here for testing. This is just like the triplets you told us about."

"Anyone else have a thought?" Dr. Langley asked as nearly every hand in the room went up.

"Even though one of the twins was raised in a typical blue collar family and the other in an upper-middle-class home, they had the same level of education, favored the same subjects in school, and worked at very similar jobs. One was a tax accountant, and the other an auditor," Jasper responded with amazement. "Heck, their wives even had the same first name."

"You've all come to the same astonished observations that I know Dr. Bouchard did when he began this study. Now in this first study, these two gentlemen, along with around a dozen more, completed a battery of tests at our university — the Weschler Adult Intelligence Scale, the Minnesota Multiphasic Personality Inventory, the Strong Vocational Interest Inventory, and several other tests of personality traits, along with personal interviews. The degree to which they manifested similarity on an array of characteristics was indeed remarkable. Yes, Sondra, you have a question?"

"Aside from the two men we saw on the video, how do we know that the others did not share some similarities in exactly how they were raised, even though they were raised apart?"

"That is a very good question. There is what is known as a socio-economic rating scale that gets at the degree to which the rearing environment is more or less the same. It looks at household income, family living arrangements, and other socio-economic factors. Dr. Bouchard used this scale to try to get at … um … to statistically adjust for those factors. In other twins studies since Bouchard, using parenting style indicators, Bouchard's findings have been confirmed. Although the sample size in Bouchard's study was not extensive, his data show that even when comparing twins reared apart in dramatically different home environments, there was remarkable concordance, particularly in the domain of intelligence. Clearly, something genetic is at work. Since this study, there have been dozens of other studies that, as I say, have replicated his findings.

"Before we go on, I'd like to review a few points about intelligence, since we talk so much about it in this class. It's important to your understanding of where we are going, and for some will no doubt be controversial.

"Generally speaking, intelligence is defined as the ability to learn, understand, solve problems, and adapt to new or challenging situations. It has to do with the ability to apply knowledge to manipulate one's environment, and to think abstractly. It is what we mean by general cognitive ability, or simply 'g', and it is measured by intelligence tests. While IQ test instruments measure a variety of abilities such as numeracy, spatial relations, verbal abilities, and the like, it tends to be true that those who excel in one subtest also tend to excel in all, hence the concept of 'g,' or general intelligence. This idea remains somewhat controversial, although most in cognitive science agree.

"As you can see on this slide, test scores for 'g' are presented in a histogram with a scale where 100 represents the mean, and scores above and below 100 form the shape of the familiar bell curve. A little over two-thirds of a population sample will have IQs in the range of plus or minus one standard deviation, or on our scale from 85 to 115. IQs from 115 to 130 are considered above average, a range within which most college students fall. This range accounts for about 14% of the population. IQs above 130 are considered gifted, and persons with IQs below 70 are generally considered to be cognitively impaired. Each of these latter two groups, representing the extreme ends of the IQ scale, account for around 2% of the population.

"IQ scores can first be reliably measured around age six, are generally stable over one's lifetime, peak around age 25, and trail off in later years. IQ is a fair predictor of years of education, income, health, socio-economic status, and social adjustment, though not necessarily happiness. In general, however, we can say that a high IQ confers advantage with respect to life outcomes. There are known IQ differences between population groups. IQs are a function of genes, shared environments, and unshared environments. IQ is not fixed, though as I say, it is quite stable. That said, IQs have moved up for large population samples over the years, though this effect, known as the Flynn Effect, seems to have diminished in recent years for more industrialized societies." When Ann finished, almost every hand shot up.

"Again, as with an earlier lecture, I seem to have struck a nerve," she said with a smile. "This no doubt will take the better part of our remaining time, so let's proceed," she said, pointing to a student in the front row. For over thirty minutes, she responded patiently to misconceptions, emotional arguments, and demands for proof. With only five minutes remaining in the period, she had to close off the discussion and offered to meet with anyone who wished to discuss these claims further.

"Listen, I know you want to more talk about this, which we can do briefly next period, but I think we've got to wrap it up here. I'd like you to read chapters eight, nine, and ten for next week. We'll be delving into how we measure heritability, and the degree to which shared and unshared environments play a role."

Out of the corner of her eye, she saw that Chuck Nathan's friend had his hand raised. She turned away quickly, but it was too late; he saw her looking at him and began. "If IQ is so important, does this mean that we should be figuring out how to make some kind of baby geniuses?" Only a few of the students were still in the classroom at this point. They paused to hear her response.

"Baby geniuses ... designer babies, as they are sometime called ... are at present a scientific impossibility. Having said that, it might at least be possible one day to have a system to screen for general cognitive ability prior to an *in vitro* birth. In any event, I see our time is up. I have to rush off. I'm sorry; I hope you will excuse me."

Ann looked up from gathering her belongings and saw David staring back at her from a seat in the back row. A smile creased his face. He got up and walked toward her. Before he made it halfway across the room, she said, "Oh, David, I'm glad you're still here. I wonder if you have time to discuss something with me in my office a bit later. I've got to run a quick errand now, but I can meet you there in an hour, if that's okay."

"Sure, Dr. Langley, I'll see you there.

CHAPTER V

ANN GRABBED HER BAG and dashed out to the small faculty parking lot behind her building. Intent on completing an errand in time to meet with David, she nearly knocked over a student while stepping off the curb. She excused herself, and then recognized the student as Chuck Nathan's friend. "Oh, pardon me," she said, now more flustered. "It's ... um, ah, Rick, from class. Right? I'm so very sorry. I guess I wasn't paying attention. Anyway, I hope you'll excuse me."

"No problem," he said as he started to walk away.

Feeling bad, she added, "What brings you through this back alley? I thought this was a place only for professors ... to make a quick getaway, you know." Her attempt at humor was lost on him.

"Is that your car?" he asked, turning back to respond.

"Why, yes," she said, trying harder to sound friendly. "Why do you ask?"

"It looks like you've got a tire low on air," he replied. He walked to her car, knelt down and ran his hand around the tire. "It's leaking air now. I can feel it. If you've got a spare, I can change it for you."

"Oh, no, that isn't necessary," she answered. "You've obviously got some place to be. I wouldn't want to make you late," she said. "I can just ...."

"It's no problem. It'll only take me a second. My job can wait," he said confidently.

"You sure? I wouldn't want you to get in trouble at work."

"Like I said, it'll only take a minute."

She pulled out her keys and opened her trunk. Rick gathered everything he needed, and in less than ten minutes the job was finished.

She rummaged around in her bag for her wallet, and extended a twenty to him. "I'm afraid that's all I've got now," she said. "I can give you more in class. I really do appreciate it. I can't thank you enough, Rick."

"It's okay, Dr. Langley. I appreciate the offer. I'm just happy I could be of help." Rick thanked her again for the offer, grabbed his backpack and dashed off to his job.

By the time Ann got back to her office, David was waiting at the door. Still frazzled by the delay caused by the flat, she fumbled to get her key in the door. Turning to David she asked, "Say, do you know Rick — I think his last name is Salvage, or something like that?" David waited for her to unlock the door before answering.

"Rick Selvaggio? Yes, I do. He and I worked together at a restaurant in Minneapolis last summer. He was a line cook. I was a waiter. I've never seen anyone work as hard. Why do you ask?"

"Hard worker?" she reflected, almost as if she were talking to herself. "I mean, that's interesting. Anyway, I was just wondering. Just a short while ago I ran into him in the parking lot behind the building. He fixed my flat tire."

"I know you must think he's a bit of a screw-up," David said.

"Well, he's said …."

"He's said some stupid things in class, I know. That's just his sense of humor. He was like that at work. It's what got him a general discharge from the Army. But he's not a bad guy. Like I said, he may not be a sterling student, but he works extremely hard, and I've never seen a guy who could keep so many orders straight while working over a hot grill. I think he's working afternoons and evenings at a burger joint in Dinkytown."

"Full time?" She paused a moment to reflect on what she'd just heard before putting her bag down amidst the pile of papers on her desk.

"Why don't we sit over there at the table?" she said. "If you've got the time, I'd like to hear a little bit about what you're learning from all the reading you've been doing. The research papers, I mean. I also wanted to take some time to go into our research in a bit more depth. I think it will help you as you begin to outline a plan for our database. Oh, before I go on, I think we should keep the conversation about Rick to ourselves. As a rule, I don't like to talk about students unless it's academically necessary. It's just that — I think you can see the reason for my interest. Anyway."

"I understand," David replied. "I'll say nothing of it."

"I'm curious. Do you work also, besides the job in our lab, I mean?"

"I do," he said. "I work maybe ten to fifteen hours a week for my uncle. He's a builder, and I do some errands for him … get building permits, and things like that."

"Your grades don't seem to suffer for it. Anyway, that's enough of that," she said. "Let's talk about the research. Tell me, what have you been reading, and more importantly, what have you learned?"

"Oh, gosh. Where to begin?" David tilted his head back and rolled his eyes upward to gather his thoughts. "Well, there have been several reports on, ah … in one case on synaptic pruning in the cerebral cortex, and another on the migration of nascent neuronal stem cells to the cerebral cortex during early brain development. I didn't completely understand it all, but did get a sense of what's being accomplished. I suppose I should also mention that several GWAS reports besides ours show we are reaching the point of diminishing returns, at least insofar as accounting for a reasonably high proportion of the variance in intelligence attributable to genes. Our work was referred to in those papers. I could go on. Oh, I forgot. There were two other GWAS studies that looked at genetic variance of individuals with extremely high measured intelligence. These were kind of cool because they looked at IQs of 170 and above. I think it was the gene ADAM12 that popped out of both studies. Curiously, one of those implicated ADAM12 as affecting survivability of several varieties of cancer … ovarian cancer being one."

"Well, you certainly have been busy!" Ann remarked. "The studies you cite represent quite a spectrum. Therein lies the problem, as I'm sure you can see. But let me back up a minute and talk about what our consortium is focusing on." Ann gathered herself, and in doing so assumed a posture that conveyed to David that he was in for another of Ann's expositions on brain science. "So, maybe to begin at the beginning. Since the beginning of scientific inquiry into brain science, researchers and philosophers have haggled, often quite emotionally, about whether it is nature or nurture that gives rise to human variation. In successive waves, one side or the other has yielded ground. At the same time, neither side has achieved any kind of compelling scientific evidence for the dominance of their argument. To be sure, both arguments have validity. Let me just say that with respect to environmental influences, as we will soon be discussing in class, there are two types: shared environment, and unshared environment. In the former, we may share certain external influences, whether from our culture, our subculture, or our family. As to the unshared environment, we have our own friends and unique life experiences. The way in which we respond to those external influences is unique to us as individuals. It can also be said that our genes partly guide what experiences we select for ourselves, and how we respond to those external forces. So genetics are still involved, even through the unshared environment.

"Still, skeptics remain about the predominance of genetic influences on behavior. For many people, the consequences of accepting even some smaller level of genetic influence amounts to biological determinism. To them that is frightening. Some even dispute that there are inherent differences in general native intelligence. Those who oppose the idea do so with vehemence, and prefer to think of humans, save for those with rare genetic mutations, as being a clean slate, influenced predominantly by their environment. In all this debate, they see a classic struggle between the social classes.

"Our aim is not to settle the argument. It's simply to understand the mechanisms of intelligence. We leave the debate for others. We recognize that our work is controversial. Therefore, we have to be very careful about how we present our findings. The brain is the very seat of humanity, so it is natural that we should wish to discover how our brains work. These aren't just biological questions, they are moral as well; however, our focus is on the biological. In the process, we want to determine just what it is that makes one person's brain more efficient and effective than another's. As I've said before, intelligence is advantage. I do not really see that as subject to debate. The real question, then, is: What is it predominantly that provides that advantage?

"Contrary to what many fear, we aren't about making designer babies. With over twelve hundred genes contributing in some small way to intelligence, it is unlikely, I'd say even impossible, at least for now, to think about how we might edit all those genes, and at just the right time, to make a more intelligent person. You can see from our own attempt to refine a polygenic scoring system that we cannot yet accurately predict intelligence from our genes, but we are making progress. Right now, though, we can do just about as well predicting IQ if you tell me the family's income and the mother's level of education.

"All that said, there probably is some genetic mechanism that we don't yet understand that will help close the gap. We're just not there yet. As you can see, the many brain research projects going on now, whether analysis of protein expression patterns, neuronal connections, gene editing, brain scanning or cell mapping, are getting us closer to that point. Work in mathematical modeling and AI is also moving us closer to understanding how our brains work, so you can see how important it is to share with other researchers what we learn.

"Oh, and just one last thing. There are those, particularly those with the power and resources to do great harm, who will not abide by convention, or who will act without a sense of common morality. They will push ahead of the science, just as they always have. Seeking advantage — it's a blessing and a curse, I think. Though it moves science and technology forward, it also has the power to move society backward. I confess, we all seek advantage in one way or another. I shudder to think of it, but even now there are bio-hackers out there, clubs of them, who can buy materials online to run their own gene editing experiments. It's now possible for a relatively unaccomplished bio-hacker, maybe even a high school kid, to sequence and alter genes. They can buy all they need with a few mouse clicks. We need to be on our guard.

I would caution you: be alert to anything you might see or hear of that exceeds agreed-upon genetic research standards ... particularly work that alters the germ line. "

Ann brushed the hair back from her forehead and leaned back in her chair to signal she was through.

"I understand. I've got just one question. If I'm right that work on developing a polygenic scoring system is reaching the point of diminishing returns, at least based on what we know now, does that mean our funding is in jeopardy?" he asked.

"First, let me say that your job is not in jeopardy. Your position, for now, is funded by the university. However, the funding of our GWAS study, at least as previously defined, is coming to an end. When we finish our final report and submit it to several journals, we will have completed the work product for which we have received those particular grants. That in turn will affect only a few of our post docs, who no doubt will seek research opportunities elsewhere. Such is the life of a post doc. Our doctoral candidates will continue in the lab under my supervision, and work on their own particular interests. For most, that means perhaps segmenting our GWAS database to look at connections between intelligence and some particular subgroup. As with the study you cited that looked at very high-IQ persons, some will evaluate correlations between those at one standard deviation above the mean to those one standard deviation below. Others might look at those with especially high subtest scores compared to the overall group average, and so on. The details of those research efforts are yet to be finalized. As I say, though, for now your job will remain the same."

"Thank you," David said. "I'm still very excited to be working with you and your team, and look forward to helping any way I can."

"Great. That reminds me. Let me show you something really cool, a software application from Berkeley." Ann pivoted toward her desk and tapped her keyboard. "Take a look at this," she said proudly. "It'll blow you away." David scooted his chair close to better view a high definition 3D image of a brain. Ann rotated the image, naming prominent brain structures as she went. She then scrolled to look deeper inside the brain, again naming more features. She changed the image to successively higher levels of resolution, and showed axon connections between brain structures. Finally, she showed a detailed cellular map of the brain that she could zoom right down to an individual cell.

"Gosh, this is amazing!" David said. His response revealed more emotion than she had ever seen from him.

"Isn't it?" she replied, likewise enthusiastic. "Very soon I expect we'll have a version that can show neurons firing in real time during performance of some complex task. I also expect it to show gene expression in individual cells. This is done with functional MRI, single-proton emission computed tomography, diffusion tensor imaging, and the like. What's new is the ability to use a brain cell atlas and simulate cell specific gene expression in real time. We'll be able to see all the cells that express the same gene in different brain regions, or to look at expression only in specific brain regions. While this is only a computer model of a real brain, it should give us new insights on how and where gene expression works in the brain during selected activities. In time, I think you can see how we might use this to model both highly functioning brains as well as those with cognitive limitations."

David smiled broadly. She said, "I knew you'd like this. I can't wait for us to work on answering some of the big questions together. For now, let's just leave it here. I've got to get ready for my next class. In the meantime, why don't you and Michael meet and put down some preliminary thoughts? As a first step, maybe you can consider a broad outline of how we might sort research into categories. We can all meet … umm … how about Wednesday, say 4:00?"

"Can we make a little earlier?" David said. "I've got a quick thing to do for my uncle."

"Oh, that's right. You've got another part-time job," Ann said, laughing in amazement that he could balance so many different activities. "Sure, that'll work. See you then."

◆

One day after giving a short quiz, Dr. Langley passed the corrected papers back. On Rick's paper, she had written a note for him to stop by her office immediately after class.

"Rick, thank you for coming by. I hope my note didn't alarm you. First, let me thank you again for helping me out with my car. I'm afraid that when it comes to mechanical things, I'm a total klutz," she said.

"Oh, no problem, Dr. Langley. Glad I could help."

"I know you've probably got to get going, so I'll get to the point. Rick, you're failing my class, and I'm concerned for you. We're far enough into the semester that it probably wouldn't do you any good to drop the class, at least as far as a tuition refund. I was just wondering why you chose this class. Does it fulfill some requirement for your major? And more important, is there anything I can do to help you?"

"Well, Dr. Langley," Rick said, as he rested his elbows on his knees and looked down at the floor, "actually, I'm not completely sure. I'm majoring in entrepreneurship. I realize this class isn't part of the requirements. Fact is, I've never really had anything beyond high school biology, so …."

"You know that Genetics 101 and Psych 101 and 102 are prerequisites," she answered. "I'm wondering how you got in ... how the system didn't catch this."

"I'm not sure," he said. "I saw the blurb on it, and was curious, I guess. Just trying to figure stuff out."

"Figure stuff out? Like what?"

"Well, it's kind of a long story, but I'm, ah ... I was just trying to understand a bit why I, ah ... why certain things have happened," he said.

Ann looked at him for a long time without saying anything. She resisted a surge of compassion, trying to remain professionally detached.

"There were some things growing up. They weren't all that great — well, anyway, I'm just not sure I want to get into it. I guess I was just curious, that's all."

"Rick, I don't know if my class will help with that, but you are sure welcome to stick it out to see if you can satisfy your curiosity. Though I doubt we'll get into enough detail for that. Anyway, if there is something I can do to help you, you know my office hours. I'm almost always available, and will help in any way I can. I could maybe suggest a tutor for you if you think that will help. Maybe one of our grad students is available. Would you'd like me to look into it for you?"

"That's all right, Dr. Langley. I'll just have to try harder, I guess. I understand it if you have to fail me. It's not your fault. You're a good teacher and all. I suppose I don't have the background for any of this, but it's interesting to me anyway. I guess I'll just stick it out ... that is, if it's okay with you," he said.

"That's just fine, Rick. I understand, and good luck to you. But please let me know if there's anything I can do."

After class, Lilith stopped by Ann's office. Her timing was impeccable, as usual. Perhaps she had a sixth sense about the onset of another of Ann's inner struggles. When she slipped in the partially open door to Ann's office, she found her friend staring out the window. "What's up, Dr. Langley? Another problem with someone's sophomoric behavior?"

"Oh, hey, Lilith. I didn't hear you come in. Sorry, I was just thinking."

"When did that ever do anyone any good? If you're free, I thought I might drop by this evening. We can catch up. I'll pick up a bottle of Prosecco."

" Great! I'd welcome the company. How about six? I'll stop and get us something from the deli counter," Ann replied, now more fully engaged.

"I'll see you then," Lilith said, quickly starting out the door. "Oops, I better get going; there's an angry mob of students headed this way, and they're all wearing black bandanas over their faces."

"Lilith?"

"What?"

"You're full of it!" Ann declared. "It's why I love you."

Lilith raised one eyebrow. "I'll see you at six."

Ann rummaged absentmindedly through the pile of papers on her desk, and then frustrated by her inability to focus, flipped on her computer. She hadn't gotten past logging in before she found herself staring blankly out the window again. Her encounters with Rick jarred something loose. She wasn't sure what to make of it, but it caused her to reflect on his story. However sketchy the details, enough of a pattern emerged to raise questions in her mind, and it troubled her. Other than thinking about her grad students and post docs, she didn't as a rule think much about her students as individuals, at least as complete persons. They were just students among a passing parade. Sure, there had been the occasional student who stood out from time to time. Those were the ones she got to know. Those who wanted to move in a different direction academically were dropped from her thought life. Teaching three sections a week for more than ten years in a large research university will do that to you, she thought. At first, as an assistant professor, she had considered reaching out to students. She had tried to get to know many of them personally. The demands of her research, as well as the practical reality of participating in the individual lives of her students had proved unrealistic. She had learned that pouring herself into her lectures and defining a groundbreaking path for her own research were the best ways to have an impact. Ultimately, she had learned that students had to take responsibility for their own learning, and she would just concentrate on her teaching. The metamorphosis gathered speed in her early years when she had begun to notice many of her peers flame out. Some had failed to attain tenure, or just changed careers. She realized then that she wasn't going to make it in academia if she placed the advancement of others ahead of herself. You didn't have to read too many biographies to know that the ones who truly made a difference devoted themselves to gaining advantage, whether through sheer will power or developing the exceptional gifts they had been born with. Knowing this, then, it was peculiar that Ann would have devoted even a small part of her thought life to Rick.

CHAPTER VI

ANN WIPED HER HANDS on a dish towel and scurried to the door. With flagrant disregard for propriety, Lilith burst across the threshold, arms flung wide. In an ensemble more suited to Miami Beach than a staid Midwestern city of Lutheran heritage, she greeted her hostess and friend with an effusive embrace. Believing no lily could be too gilded, she wore a low-cut leopard print jumpsuit with a Frederick's of Hollywood-style push-up bra, accessorized with a wide-brimmed straw hat and sunglasses the size of two small television screens. She planted a scarlet kiss on Ann's cheek, made an imperfect attempt to wipe away the sanguinary smudge, then brushed by her friend to deposit two bottles of wine in the fridge. In the reverie of another Friday night emancipation, the two friends required neither the support nor the companionship of men. Perhaps it had been the routine of unremarkable days, the drudgery in all things academic that had led them to such unreserved silliness. It was to be yet one more occasion for sharing pent-up desires and the many frustrations from which no one is immune. In the confines of Ann's apartment, they were free; nay, they were bound to act to the solace of the other. Wine was uncorked, glasses filled, and in a simultaneous toast to the "life-giving properties of Prosecco," Ann and Lilith hastened on to the satisfying first sip of bubbly.

"Why don't we sit over there?" Ann gestured toward the sectional, as she opened the oven door to freshly toasted crostini. "Give me a minute, and I'll be right with you. I'm so anxious to catch up on what's gone on since our last date."

"Likewise. Oh, speaking of dates, have anything new to report?" Lilith inquired, hoping for some tidbit of delicious gossip.

Setting the plate of freshly toasted tomato-topped bread on the coffee table, Ann joined Lilith on the sofa. Before she'd settled in, her dismissive huff conveyed all that really need have been said in reply to Lilith's question, but she went on, "Nothing to report here. How about you? How have you been keeping yourself busy these last couple of weekends? Oh, before I forget, allow me to comment on your … well, your … how shall I say … most alluring outfit. It's simply marvelous, my darling. Where exactly did you find it?"

"I'll get to that, but first I have something for you." She reached into a deep pocket and pulled out two black balaclavas. Donning one, while handing the other to Ann, she said, "Like I said before, if you can't lick 'em, join 'em," at which they both giggled.

Feigning the pose of an ANTIFA punk, Ann raised a fist and yelled, "We're shuttin' 'er down, man. We're gonna start a revolution, you fascists."

"Every time I see one of those guys, I can't help but think of that scene from *The Big Lebowski*, you know, where the leader says, 'Where's the money, Lebowski. Where's the money? ' Then he tosses his ferret into the bathtub between Lebowski's legs." Lilith had the imitation of the nihilists down cold, not surprisingly; it was one of her favorite movies. She'd used the film in her classes to teach concepts of Plato and Nietzche to great effect.

The two friends riffed on scenes from the movie until they laughed so hysterically that Lilith had to excuse herself to use the restroom. When she returned, she said, "But to answer your earlier question, about the outfit, that is, it was a gift from Peter, er ... you know him as André. We're going to one of those all-inclusive resorts in Jamaica over winter break. I guess he thought it would be fitting for me to have something suitable for the occasion."

"Peter? André? Girl, you've been hiding things from me. Come on, out with it."

"André is like a stage name. He says it gets him more tips. What we saw that night at the Dakota is an act. Well, sort of. He really is quite the charmer ... and a great lover, if I do say so myself. He's got a marvelous sense of humor, and is thoughtful and caring. If not well read, he more than compensates in other ways, if you know what I mean."

"He makes you happy, then?" Ann asked in all seriousness.

"He does," Lilith replied defensively. "I know you may find that surprising, maybe even a bit amusing, but he really does. Look, a girl doesn't need a college professor type to be happy. In fact, I'd sooner die an old maid than date a professor. You have to admit, he is good-looking. Besides, he's kind, and he's emotionally generous in a way that no man has ever been with me. What could be wrong with that?" Lilith responded.

"I didn't mean to sound critical. I'm not. His qualities, as you describe them, are what any woman would want. It's just that, well, he just didn't seem your type, that's all ... at least the first time we saw him. I'm sorry for giving you a different impression, and I'm sorry for misjudging him. I seem to have been doing a lot of that lately ... misjudging, I mean. Please forgive me. Hey, excuse me a second, I've just got to check on something in the kitchen." Ann was glad for the opportunity to create some space between herself and Lilith. She realized that her own unfiltered observation had touched a nerve.

"Don't worry about it, love. No one is more aware of my poor choices in men than I am. It's just that, honestly, for the first time in my sordid dating history, I can say a man truly makes me happy. Besides, smarts aren't the only thing that matters. As you know, I've got enough for both of us."

Ann came back to the couch with the opened bottle of wine and topped off both of their glasses. "You know, Lilith, as I said, I've been guilty of making other misjudgments about people of late. I've got this kid … you remember me mentioning the troublesome student I've got in my behavioral genetics course … not the one who went to the department chair, but the one who made all of the inappropriate comments that set off the brouhaha? It turns out that there's more there than meets the eye. Oh, he does really sometimes say some stupid things … really stupid things. But he's got a story that I couldn't have appreciated until I ran into him in the parking lot behind our building one day. Turns out he noticed I had a flat tire, and made himself late for work in order to fix it for me. In the course of chatting with him, and later talking about him with another student, I learned a bit more about him … enough, anyway. Turns out he's working full time — nights and evenings — to put himself through school. It's maybe not so surprising that he's been struggling in class. I recently called him into my office to discuss the fact that he's failing. I offered to help in any way I could. He shared enough with me … let's just say he implied that he's got some family history. Perhaps that's what's gotten in the way of his social development, not to mention academics. It's true enough that he probably never was going to be academically successful, but he is a hard worker, and considerate. Honestly, he has been a pain in the ass in class, but given where he comes from, well … I guess I have to forgive him, at least a little. Like you said, maybe smarts aren't everything."

"Sounds like an interesting kid. He's got a story … just like everyone else. Maybe school isn't his thing, but then it isn't for a lot of people. So what? Intelligence may be overrated. Like my father use to say, 'The world is really run by 'C' students.' But then again, you and I both know it's really run by women." Lilith gave a little shimmy to punctuate her claim.

"Well, I'm not sure about that, but it is true that the great mass of people occupy the middle part of the bell curve. Those at the high end, women or not, are the ones that make the great leaps forward. Intelligence may not be the only thing, but it is important."

"They move the world backward, too," Lilith replied as she stood up and looked toward the kitchen. "Is that steak I smell?"

"Oh, I'm sorry. Yes, it is. We're having a steak salad. I grilled a marinated flank steak earlier. I was just warming it. I've got a gorgeous baguette, and of course we've got another bottle of Prosecco. You ready?"

"Absolutely," replied Lilith. "You can catch me up on everything else in your life besides men."

"I've got to ask, Lil, why so much interest in my work? I mean seriously."

"It's the brain, hon. You've said so yourself. The brain's the key to everything. What we think, and how well we think and feel about things is at the root of everything that is right and wrong in the world. From what you've been telling me, genetics is a big part of that. So why the heck wouldn't I be interested?"

"Fair enough. I can bring you up to date, but let's eat first."

The two chatted on lighter topics over dinner, nearly finishing off the second bottle of wine before adjourning to the couch for dessert.

Ann brought out a chilled bottle of vinho verde to accompany fruit tarts that she'd purchased on her way home.

"That was a wonderful meal," Lilith said graciously. "Thank you, and thank you for the laughter. What a great way to end the week." She raised her glass to toast her hostess.

"You're very welcome. Anytime," Ann replied, raising her own glass in response.

"Now on to more serious business," Lilith said. "What have you discovered about the brain lately?"

"Well, one thing I discovered is some fresh talent… in fact, I hired someone whom I learned had a deep interest in our work. He's one of my most promising students. I brought him on to help us ramp up the use of artificial intelligence in our research. You ran into him during the first week of the semester. He dropped by my office that time we made plans to go to the Dakota."

"I don't … well, on second thought, was he the one wearing a baseball hat?"

"No, he was the good-looking young man that popped in just as you were leaving," Ann replied.

"Oh, that one. I saw him coming down the hall toward your office. He was good-looking. I trust he has more going for himself than that, though."

"Indeed. He's pretty insightful. Maybe a bit too serious, though. Anyway, I hope he'll loosen up. He's entered a doctoral program in behavioral genetics. I'm going to be his academic advisor. I expect he'll be quite an addition."

"So, then, what has your team of brilliant researchers been finding as far as being able to predict intelligence from someone's genetic profile?" Lilith inquired.

"To tell you the truth, we're kind of at a dead end. We've been finalizing our report, and we're expected to submit it to several journals in the coming weeks. As I've mentioned before, we expanded our sample size and did identify more genes that vary with intelligence. But we still haven't managed to account for the level of variance that we expected. To top it off, a research team in Holland has surpassed our gene count slightly. They just announced it last week. It looks like if we were able to continue, we might find a few more relevant genes, but still wouldn't account for anywhere near half the variance in IQ. So, to sum it all up … we're at the end, and we came up short. These studies are enormously expensive and time-consuming, so until we get a better handle on all of the biological processes that influence brain function, it seems like our resources might be put to better uses."

"And what might those be?" Lilith asked.

"I'm not sure. Our grant money is nearly gone, so we're in the process of hunting for more. It's not so easy to find. For now, we'll be able to keep a few of our doctoral students busy doing work on smaller studies — for example, we might look at genes for savant-like abilities in math, languages, even arts. Compare those to average folks."

"So, does this mean you're discouraged?" Lilith sounded deflated as she reached for the wine bottle.

"Yes, but I do think we have a lot more to learn. One day I expect we'll make advances on how to treat complex brain disorders, but like intelligence, those are multi-faceted problems … and we're just not there yet. We've begun to disperse the fog somewhat. It's just that we don't have a way to develop a very accurate polygenic scoring system.

It looks like the best we can do, at least in terms of accurately predicting intelligence at the embryonic stage, is to just measure the intelligence of parents and grandparents. We can get a general idea from embryonic DNA, but that's only a guess. The best approach is not so different from what the early eugenicists proposed. You've got to marry well if you're looking to have a smart kid."

"I've noticed you keep bringing up babies. Had more thoughts about the subject?" Lilith asked with a suspicious smile.

Ann puffed through pursed lips, and then took a deep swig of wine. "No chance," she said. "It's just that whenever I talk about my work, someone invariably brings up designer babies. And speaking of designer babies, did you ever wonder how these Hollywood types, married in their late forties … even into their fifties, manage to have so many twins? Most of the time it's one of each, and never do we hear about the birth of a kid with defects. Well, I can help you. They do *in vitro* fertilization and use DNA sequencing to accomplish what's called pre-implantation genetic diagnosis. They cull the bad embryos. There's your designer babies." Ann was beginning to slur her words slightly.

"I guess I never thought about it, but come to think of it, you're right. God, we're a mess, aren't we? Humans, I mean."

"Not only that, but we've even been able to take skin cells, turn them into what's called pluripotent stem cells, and then tease them into sperm and eggs, use *in vitro* fertilization and produce babies. We've only done that with mice so far, but it's just a matter of time before someone somewhere in the world tries to do it with humans. You wouldn't even need a man to have a baby." Sufficiently tipsy, Ann was beginning to get more agitated than usual.

"Not my cup of tea, but hey, girl, I've got to say you're getting a little worked up here. Why don't you have another while I go to the ladies' room?" Lilith got up, stumbled, and then turned to look sardonically at Ann; "I don't know about you, but baby or no, I've got to have me a man."

When she returned, Ann turned on a selection of jazz and the two slumped more deeply into the couch for some mellowing time. After about fifteen minutes, Ann got up and offered to split the last tart. In the waning hours of the night, they chatted aimlessly until Lilith began to nod off. Around midnight, Ann got up, jostled her friend and offered the day bed in the spare room. Ann laid out a robe, towels and a pair of too-small pajamas, and bade Lilith good night.

◆

"Good morning," Ann said. "Did you sleep well?"

"Like a baby. Must have been something I drank," Lilith answered. "Something smells scrumptious. You must have been up for a time."

"I've got some muffins going. Help yourself to the coffee."

Lilith browsed the selection of coffee pods and inserted a dark roast in the maker. "What have you got going today?"

"Oh, nothing special. I've got to run over to the lab and check on some things. After that, probably just lay low. How about you?" Ann asked.

"André … er … Peter and I are going to a concert tonight. We'll grab a quick bite, then head to the Target Center."

"Flying Pigs?"

"Sister, you've been stuck in the lab way too long. That group is so yesterday. Nope, it's The Klingons — an alt-punk group out of Seattle. Peter says they're the best."

"Alt-punk! I have been cooped up too long. Anyway, I don't mean to change the subject, but I've been meaning to ask why we always talk about my work. Sometime I'd like to hear more about yours. It seems like I always monopolize our conversations, and think it all must be a little boring for you," Ann said.

"Not really. I am interested. There's really not much to say about what I do. I teach. There's no new research in my field … nothing earth shattering anyway. All the great thinking has been done. Only thing left is to comment," Lilith said, sitting down at the counter.

"No one comments better than you. That last piece … what was it … *Nietsche and the Big Lebowski, A Tale of Two Slackers*. It was priceless!"

"Publish or perish, right? A girl's got to do what a girl's got to do. You know, over in the Philosophy department there just isn't much competition. All I have to do is teach students to think, and maybe look at the truth in different ways. For us, the truth has aspect, you know what I mean; it has facets. We examine truth from every angle, and then never reach any conclusion. We study the forest. You all look at the trees. You try to figure out how they work, maybe breed better trees. We try to figure out what the forest means. Your students solve problems. Mine, if I'm lucky, just try to think about problems in a different way. I suppose I'm a little too contented. I never was very competitive. Competitive advantage has no importance to philosophers, at least this one. Whatever anyone does in the sciences lately, always seems to focus on making things better than they naturally are."

"I know what you're saying." Ann said, as she opened the oven and pulled out the tin of muffins. "I've been thinking about the same thing lately. I'm not sure why. It's true that the nature of what we do is competitive, and that we use all our resources, especially our intellect, to fix things, to make them better. We have to in order to fund our research. Admittedly the competitiveness is also to feed our pride, but you've got to admit, science has discovered a lot of wonderful things. For whatever misguided direction one wayward person chooses with science, there are dozens of beneficial applications. I guess if we have to endure someone's pride in the process, so be it. Here, have a muffin."

"I detect a bit of frustration with your work from the way you talked last night. Anyway, you seemed to suggest you're maybe reaching the end of the road," Lilith observed.

"Finding a thousand or so gene markers for intelligence is a good thing, no matter who gets the credit. But it still doesn't account for even half the difference in intelligence. We can't yet explain how it all works. I'm afraid our approach is not adding much more to our understanding. It's kind of like we need to discover a master switch, some mechanism that makes the brain work well in the case of really smart people."

"A God Particle, as it were," Lilith said.

"Yeah, something like that," Ann responded. "How's the muffin? More coffee?"

" Mmmm. Coffee, no thank you. I'd better jump in the shower. I need to run some errands this morning. Hey, listen, last night was fun. We've got to do this again soon. It's good for the soul. You in?" Lilith said, standing.

"Sure, but next time I'll wear a costume."

"Deal," Lilith said over her shoulder as she made her way to the hall bathroom.

Despite the beautiful early fall weather, Ann hesitated to dress and get on with her day. Lost in thought, she lounged on the couch until nearly noon. There was something about her encounters with Lilith that arrested her forward momentum, and caused her to reflect on her life and her work. Perhaps it was Lilith's disruptive flair, or the challenge posed by Lilith's observations that caused her to reconsider her assumptions. Their conversation had drifted into unsettling questions of advantage-seeking and competitiveness. By the time they met, Ann had already reached the conclusion that her research was finished. Though she had long hoped to make some kind of breakthrough, she was forced to set aside her pride and acknowledge that her own efforts had failed to break much new ground. Feeling diminished, if only momentarily, she rationalized that at least she had added to the list of genes that affected intelligence. That was something. It might prove useful later.

These thoughts once again triggered the guilt she felt over the failed relationship with her former husband. It had started well, but in a few short years, whether something inherent in each of them, or the demands of their work, tore them apart. It wasn't just the disheartening findings from her research that ate at her. She knew that she could somehow dress up her findings to make it seem to others that important advances had been made. She was not fooled, though. It was also the academic environment, the students, the culture, the whole thing. She really did begin to feel an impulse for change, a freshening was how she'd come to think of it. Perhaps, she had to admit, it was a craving for a deeper relationship, something more than work, and even friends. Maybe a man, but she just couldn't make the same mistakes as before.

An errant idea flashed in her brain. She tried to override it, but to no avail. Of late, every time she got together with Lilith, the conversation somehow got around to having a baby. Her own comments about *in vitro* fertilization, pre-implantation genetic diagnosis, and the very idea of having a baby without a man, though relevant to their discussion about science, had managed to touch on a subject that had more personal meaning for Ann. A noise outside interrupted her stream of thought. She turned to see that the soft shadows of morning had already been burned away by the sun's climb. She rose to ready herself for the new day.

Ann finished her errands and got to the lab in the early afternoon. She took a quick spin around the floor and checked in with the smaller-than-normal group still at work. David was talking with Michael. She chose not to interrupt. She went to her office to look over a budget report, and then busied herself cleaning up the stack of papers on her desk. She couldn't decide if what she was doing was unconscious stalling when a rap on the door broke the spell. David stuck his head in, and asked, "Got a minute?"

"Sure, come in," she said. "I was just doing some housekeeping."

"I saw you come in the lab. You seemed like you were busy, so I didn't want to interrupt. Sorry for not greeting you earlier. Michael and I were just discussing the database, possible architecture stuff. That's kind of what I wanted to talk to you about. Michael, I mean."

"Oh. Is everything going all right?" Ann asked. "You two making any headway?"

"Yes, but that's not it. I'm a bit hesitant to say this, but …."

"There's something else of concern, then?" Ann asked.

"To be honest, there is. When I got into the lab early this morning, I walked over to Michael's desk to see when we might talk. He wasn't aware of my presence, but I could see over his shoulder that he was typing something in Mandarin, at least I think it was Mandarin. Obviously, I couldn't read it, but he had some of our preliminary polygenic scoring data next to his computer. He looked like he was typing from that. Since you cautioned me about sharing our data until our report was finalized, I was a bit concerned, and thought I ought to check with you. I hope this doesn't seem like I'm tattling. It's just that I'm trying to understand the limits to the kind of information we can share, and with whom."

"Thank you for bringing this to my attention," Ann said. She tried to sound reassuring, like everything was normal, but had her own reasons to be concerned. She didn't want to reveal those reasons to David. "I think there's probably nothing to it, but it's good that you're sensitive to the issue. I maybe should have told you this earlier, but David's work here, in fact his whole doctoral program, was funded by the Beijing Genomics Institute, our collaborators in China. As part of our agreement, we exchange information about our research. I'm not sure exactly what he was communicating, but I will ask. Likely, it was nothing more than regular status reports that we're obliged to give them. I guess I'm not sure about why the scoring data was there by his desk, at least other than the fact that we intend for it, after it's published, to be incorporated in our database. I suppose he was just multi-tasking. Anyway, I'm glad you dropped in. I hoped you'd be in the lab today. I thought it would give me a chance to check in and see how you're doing … what kind of questions you might have."

"I appreciate that," David said. "Fact is, I have been puzzling about all this research I've been reading lately. It all confuses me some, this nature/nurture thing. I guess I'm just wondering if the way to think about our genes is that they, in some very complex way, define our individual human limits, our potential. For that matter, do they define our proclivities? What got me thinking about this was athletics … in particular athletic ability. In some ways it seems to be analogous to intelligence."

"That's an excellent thought. And what did you conclude?" Ann asked.

"Well, take endurance athletes, for example. Like long distance runners. Did you ever notice the predominance of East Africans in those events? Ethiopians and Kenyans, mostly. They dominate, and have for years. Then look at sprinters. West Africans and people from the Caribbean dominate. It looks like genetics are at work, but you can find articles that say the difference is primarily cultural, or a matter of training. There is some evidence to support both arguments. But if it's just a matter of training, you've got to think there'd be more exceptions to the rule. There's not much. These differences have been in place for many years. Interestingly, most of the elite track athletes from around the world train in the United States. If it were just a matter of training and culture, some of the different cultural influences would have been washed out.

"Then look at the difference in body types … you know, the lean and willowy versus the muscular, the respective slow twitch versus fast twitch muscle argument. Capability seems to be largely built in.

"I played halfback in football, but was always slow. No matter how much training I did, my speed never improved. Something in my genetic makeup seemed to limit me. I'm not saying that there aren't sports where practice and skill building aren't key, but running, at least at the elite level, where everyone presumably trains about the same, seems to be different. The upshot is, I'm wondering if intelligence works the same way. Do you think our genes mostly define our potential, at least when all other things are equal?"

"You certainly have been busy thinking, David," Ann said. She looked down at her phone momentarily, then turned back to David. "Whew, I just noticed that it's getting late. You hungry?"

"I haven't eaten, so yes, sure," he replied.

"Why don't I see if someone else wants to walk over by the stadium and grab a bite?" she said. "We can pick up on our conversation over a beer and some wings. I'll just pop out and see who else wants to join us. Sound good?"

"It's a plan," he said.

Most of the folks in the lab had drifted away by then, and of the few who remained, only Amy was free to join them.

"I guess it's just you, me and Amy." Ann said, stopping by David's desk after making a swing through the office. "Amy's eaten already, but said she'd join us for a beer before she has to meet up with some friends. I'll just grab my purse and meet you both by the door."

The old fire hall, now a sports bar, was packed, especially given that Minnesota was playing Iowa in Iowa City. All monitors were tuned to the game. They were seated at a small table on the patio, away from most of the other patrons. They ordered food and drinks quickly, and picked up where they left off.

"So, Amy, David was making a comparison between phenotypes, in this case athletic ability — particularly track — and intelligence, and how they are both genetically influenced. I think the comparison is fair. They're both complex traits. Still, intelligence is more complicated, I think you would agree. The genes for conformation ... I believe that's a useful word here ... body type, size, muscle development, are similarly large in number, and contribute to the performance of elite athletes, but so too does the central nervous system, and for that matter personality. In recent years, work has been done on height, muscle development, oxygen uptake, even body mass, so we're beginning to understand genetic influence on some of these traits. As with intelligence, while relevant genes have been identified, they do not account for all human variation. Given the history of doping in sports, I suspect someone, somewhere in an out-of-the-way laboratory ... maybe Russia ... will come up with genetic alterations to give athletes an advantage."

"All of which reminds me," Amy added, "there's this Argentine polo player ... a real good looking guy ... anyway, he's the top-rated polo player in the world. He says that good players are abundant, but he claims the overriding difference in polo players is the pony. All the best players agree. They say he's at the top because he has the best string of ponies in the world. It turns out his ponies are cloned. He had this one outstanding horse which, with the help of a vet at the University of Texas and backed by a wealthy oil man, has cloned animals from this one horse. Though they have slightly different markings, they are with respect to conformation, temperament, and behavior nearly identical. As you might expect, they are some of the most valuable horses in all of sports. Obviously, there's an impulse to make things better, or to seek advantage through genetic engineering."

"That's amazing. So, how long can it be before we do the same things with humans?" David asked as their wings arrived. When the server left, Ann collected her thoughts before diving in ... to both the wings and her answer.

CHAPTER VII

"IT'S NEVER BEEN OUR GOAL, but someone somewhere will try to figure out a way to use genetic tools not just to cure disease or end suffering; they will use what we learn to gain advantage. It might start in sports, and by the way, as with steroids or blood doping, it will take a while to uncover cheating. Some government will eventually experiment with engineering a group of … say, soldiers, or laborers, to make them stronger, or have more endurance. It's only a matter of time before someone figures out how to improve intelligence, or will select embryos for some specific aspect of 'g' … for instance, mathematical ability."

"You could be right, Dr. Langley." David's response suggested he wasn't quite ready to believe it.

"Look, I don't think it's even a question. We've seen the pattern before. It's maybe a ways off, but it will come soon enough. For now, though, the best way to produce the kind of person you hope to have is to just mate with persons that have the attributes you seek. It seems a bit old-fashioned, I know, but it's worked for livestock and fruit trees," she said, half jesting. "Let me correct that. The operative word is 'worked.' It worked for livestock and fruit trees. Now, all kinds of agricultural products are genetically engineered with gene editing."

The waitress came by to check on their food. Ann ordered another round just as a loud cheer went up from inside the restaurant.

Amy, who had joined them for one beer, had to excuse herself and left to catch up with friends who arrived earlier than expected.

"Hey, see you later, Amy," David said. Turning to Ann, he noted, "The Gophers must have scored. It was 27 to zip when we arrived just before half time," David observed. He overheard someone say the Gophers had gotten a field goal. "Maybe the first subjects for some genetic re-engineering ought to be our football team. But then, who am I to talk?"

Ann smiled. Looking pensive, she turned the subject in an unexpected direction. "So, David, I've got to ask, what got you into psychology in the first place, and what made you consider changing directions to pursue a doctorate in behavioral genetics?"

David seemed unsure how to answer. Given the change in tone, he started to wonder whether his professor was having second thoughts about his suitability for a job in her lab. It wasn't like him to be paranoid, but the abrupt shift in the conversation gave him pause. He knew he'd have to tread carefully.

Sensing his hesitation, Ann added, "The reason I'm asking is that I simply like to get to know my advisees … to understand their motivation and their level of commitment. The road to a terminal degree, isn't without obstacles. They can be overcome if one is determined … if one has sufficient purpose. Please don't be put off by my question. Trust me, I have no motive other than getting to know you better."

Feeling somewhat reassured, David said, "I guess I've just always been interested in what really makes us tick." Having uttered what he knew to be a vacuous response, he steadied himself for the comment that he suspected would come.

"I'm surprised at you, David. I expected more." She smiled as she peered over the top of her glasses. "I think of you as being more insightful. Please try again, only this time, tell me what really sparks your curiosity."

David took a swig of beer, wiped his mouth, and then leaned forward. "In all confidence, I have a sister who ..." he began, then hesitated.

"Trust me, David, this is all just between you and me. If we are about to begin this adventure together, one that will take four to seven years, it's a good idea that I at least get to know where you're coming from. As to your sister, or anyone else for that matter, I consider it privileged information. No pressure from me."

"My sister is on the spectrum."

"The spectrum." she replied. "I never really cared for the term. We're all on a spectrum of one sort or another, aren't we?"

"Sorry. She's autistic," he corrected himself, and then paused to take another sip of beer. "She's remarkable in so many ways, but has major difficulties in relating to people. I guess I was just trying to understand the causes," he said, now feeling a bit more comfortable in discussing something about his family that he never shared with anyone outside the family circle.

"In what ways is she remarkable?" Ann asked.

"Well, the amazing thing is she quit going to school, at least formally, when she was fifteen, because she sometimes got agitated around other people. She ended up being home-schooled for a time. What's so remarkable is that she's a math whiz. I don't mean just good; she's exceptional. She's mostly self-taught, because my mother couldn't begin to teach her the things she learned."

"Like what?" Ann asked, leaning in. Her curiosity was especially aroused at the claim.

"Oh, stuff I couldn't begin to describe. She went well past advanced calculus into number theory, abstract algebra, differential topology, logic and set theory … oh, and something called real and complex analysis. I remember some of the books she worked through. They were incomprehensible to me. I think you would say she is a math savant."

"Indeed! It's rare, but as I'm sure you know, it's not unheard of. Autism is now well known to have a genetic cause, and some as-yet-to-be-determined effect on brain morphology and chemistry. It's puzzling, then, that you would have considered psychology for a Masters rather than genetics or molecular biology. Is there anything else?"

David was not surprised that Dr. Langley picked up on the inconsistency. Perhaps, he thought, she'd noticed something in his body language. In any case, he realized that he was just going to have to trust her. "You're right," he admitted. "There's more." David looked around the patio for a moment. Everyone was too far away or too engaged in conversation to hear what he was about to say. He looked at Dr. Langley a second, gave a forced smile and began. "My father is a thoracic surgeon. He was always driven, focused, but detached. They say he was one of the best. My mother was a stay-at-home mom. She was brilliant, but devoted her life to being a mother. In some respects, I think she was more intelligent than my father. She had a degree … a couple of them, actually. She worked to put my father through school and his residency, but never worked outside the home once he began his practice, at least until they got divorced. He wanted it that way. I think maybe he was jealous of her brilliance. He was old school, I guess. Anyway, she read a lot, and could talk about anything, but she was always supportive … maybe just a little more submissive when my father was around. When I was fifteen, he moved out. He moved to California with his office administrator. My mother put on a happy face for us, got a job, even started to blossom. She did what she had to do for us, but it had to be devastating for her. In the end, she had to put my sister in a group home. Of course, the state paid for it, which was even more of a blow for my mother. Just before my senior year in high school she died of pancreatic cancer. I moved in with my uncle, the one I mentioned to you. He's my mom's older brother. Smart in his own right … a bit of a hippie. So's my aunt. They took great care of me. Gave me what I needed, but weren't rich by any means. He's built a couple of houses, and lost his shirt on those. Now he just does carpentry, some remodeling and the like. My aunt works in a clothing store, but doesn't make all that much. They're both really smart, but prefer … um, you could say an alternate lifestyle. You know what I mean? Still, they actually help me with tuition … at least with what they can. So, getting back to your question about psychology … I guess being young, I was just trying to make some sense of it all." He paused and looked up to see Ann staring back at him with forced composure. David had told his whole story only a few times before. Once he had told a girl friend, who had responded with motherly sympathy, something he'd hated. The only other time was to a psychologist that his aunt insisted he see right after his mother died. Both times had made him feel weak and dependent, so he had chosen not to share his family story with anyone else, at least until now.

"So you're borrowing money, and working some to get through school?" Ann asked.

"Like everyone else, I suppose."

"Working for me half-time, and your uncle some… you're getting by financially, I trust."

He laughed. "I sell a little blood ... and other stuff on the side, too."

"Other stuff?" she asked.

He laughed again. "Not drugs." He looked down, then back up at Ann.

"Oh," she said. "I see. And the other stuff ... how's it pay?" She laughed to mask her slight discomfort at this revelation.

"The blood ... um, not great. Helps cover my cell phone bill. The 'other stuff,' as we're calling it ... surprisingly well. They're always looking for it from virile young men, especially college students," he said with self-mockery.

Ann began to smile broadly. "Now I see why you've decided to go to grad school. It increases your value. Kind of like having a pedigree ... you'll be a luxury brand. I guess you've got to do what you've got to do."

Another cheer erupted from the patrons. Dr. Langley and David looked at each other and simultaneously remarked that the Gophers must have kicked another field goal. They finished their beers, Langley paid the check, and they left. They walked in silence for a few blocks, enjoying the cool early autumn weather and the golden late afternoon light. David spoke first, inquiring of Dr. Langley whether she thought governments would step in if public and private sector labs failed to adhere to norms regulating gene editing, especially with embryos. She said that she'd wrestled in her own mind with the issue, and concluded that it was doubtful. Congress had for many years demonstrated an inability to keep current with scientific development, and sometimes had even shown a contempt for science. As she pointed out, their ignorance of technical issues was manifest often in public statements and hearings. After many years of vitriolic debate on fetal tissue research, for example, in the end they demonstrated a reluctance to wade into such complex scientific and social policy matters. In any event, the country had been so divided for decades that Congress seemed incapable of taking a legislative position on anything controversial. Legislative action would ultimately wait until something had gone dreadfully wrong, she opined. Then, the heavy hand of government would act as a blunt instrument and impede scientific progress for years.

When Dr. Langley finished what David took to be more a minor rant than temperate reflection, she checked his reaction by asking for his opinion. Knowing that he would be treading on her strongly held view, he demurred. Realizing his reluctance, Langley said, "Conforming the world to our own view, or at least trying to, is a powerful force that rises in each of us. Forgive me. I get carried away, but I guess that's what makes resolving these issues so difficult."

Sensing the need, she shifted topics as they strode through the nearly abandoned campus. Without fear of being overheard, she probed a little more, although as delicately as she could, into the manner in which David made a little outside money.

"So, David, if you don't mind my inquiring, I'm still a little curious about that … that other job you have. How does it work, exactly?"

"If selected, and maybe only one in one hundred are — at least that's what the clinic tells me — you have to sign a contract. Just to get to that point requires completion of a long questionnaire, gathering a family history; a physical, which includes collecting a sample; a DNA screen; and some psychological testing, including an IQ test. In particular, they're interested in whether there are any genetic defects or health issues. They also want to know if you have any particular talent, like, say, musical or athletic ability."

"Of course you told them you were athletic, but slow," she said teasingly. "Was that a problem?"

"It might have been! But I made up for it with my brilliance and good looks," he replied self-disparagingly. "Ultimately, they just want to know what kind of swimmers you have, if you know what I mean. If it all checks out, you go into their database. Then, their customers … er, their patients get to select donors based on attributes they want. Oh, and I had to put together a little video biography to tell why I wanted to be a donor. It was sort of like going through the college application process. They've got this web app for their clients, like a dating app — Match or Tinder, you know — only you usually don't get to see who chooses you. Either party can choose to remain anonymous. Anyway, I'm obligated to donate twice a week for a year."

"Do you hear how many times you have been chosen?" Ann asked. She realized that continuing to ask questions was becoming intrusive, but persisted nonetheless.

"Oh, gosh, I think it's something like six or seven so far. I'm not positive, but it's something like that. Like I said, I asked not to be identified by name, or to be contacted, and I don't know the name of the recipients. If I'm right about the number, I must be getting close to the limit of what they'll allow me to donate. You know, genetic diversity and all that."

Still trying to be circumspect, Ann observed, "I didn't realize Minneapolis even had a clinic like this. I thought they were only part of very select hospitals. I guess I just never really thought about it. Do you have to travel far to donate?"

"I use one in Uptown, near where I live. I can just walk to it."

She shifted direction. "Aside from the money, which I think is a good enough reason, why do you think you decided to do this in the first place? Let me ask it another way. How does it make you feel, I mean, after the fact, after you learn that someone has chosen you?" Ann was starting to sound less clinical, or at least was trying to.

"To be honest, and I didn't initially think I'd feel this way, but it makes me feel good. Not that I was chosen, necessarily, but that I could help someone. I can only guess what parents go through when they desperately want to have children but can't. The clinics take time to try to help you understand that. I suppose they're trying to make you feel like you're making a very important difference in people's lives. Anyway, if a couple is open to using someone else's genetic material … well, it says something about their need, then, doesn't it?"

Dr. Langley saw a side of him she'd not seen before. There was tenderness, compassion, something altruistic. He was pragmatic enough to have first been drawn by the easy money. He needed it. But it was the realization that he was making a life-changing gift that also mattered to him. "You know," she said, "combining genetic material from someone other than your mate must seem awkward, maybe even unthinkable at first. I suppose it is for many, but wanting a child is a powerful motivator. I guess you could say the process is just another way of adding to the stream of humanity. Anyway, David, more power to you."

"I have to say, I am proud of what I'm doing. Oh, hey, we're here. Geez, we've been talking so much I kind of lost track of where we were."

"You headed up to the office? I've got to do a few more things before I go home," Langley said.

"No, I think I'll call it a day, but I'll be in tomorrow morning if you need me to do something," David said. "Thank you very much for lunch … and the talk. I enjoyed it."

"I'll see you tomorrow, then. And thank you, David, for a lovely afternoon. I've enjoyed our chat."

Ann unlocked her door and put her purse under the desk, then headed straight for Michael Du's desk. She was glad to see he was still there. With headphones on and a relaxed posture, he was unaware of her presence until she tapped him on the shoulder.

"Oh, sorry, Dr. Langley, I was just graphing some data for our final report. I didn't realize you were here," he said, with only the slightest hint of an accent.

"Sorry to interrupt; I know you're busy. I wanted to check in and see if you've heard anything lately from Beijing." Ann knew she shouldn't ask him directly about what David reported seeing. She didn't want to seem like she was micromanaging. He had a perfectly good reason to communicate with his Beijing peers. It had been his job to keep each of the partner labs aware of what the other was doing. Michael knew Mandarin, of course, and having once worked at the Beijing lab, could provide special insight into their research methods. Because Michael's position was funded by the Chinese government, as was his doctorate from the university, Ann had chosen to be careful what she shared with him. It was an awkward position for both parties; neither party trusted the other completely. There was interdependence, but the Chinese and American collaborators were also fierce competitors. There was too much potential for intellectual property theft.

"No," he said, "I haven't heard from them in a while. I've been so focused on getting our part of the report done, I'd even forgotten to send them our biweekly status report. Sorry about that; I was starting to do that earlier. I guess I got distracted. I hope to finish it next week, though."

"Those status reports are important, especially coming from us. We've got the lead role in communicating with all of our partners. I suppose a week's delay won't hurt. Has everyone else submitted theirs?" she asked.

"Yes, they are all in. China is the only one we've not heard from or sent one to. I promise I'll check in with them next week, and get ours out then," Michael finished. Dr. Langley nodded, and mouthed the words 'Thank you' as he put the headphones back on and resumed his work.

Langley doubted his story. She knew he was seen typing from her data into Chinese earlier that day. Whatever it was must have been sent to the Beijing Genomics Institute. They were the only collaboration partners using Mandarin. She couldn't have been certain of the contents, but it was reasonable to assume it must have been something about the most recent polygenic scoring data, none of which had been approved for release. Unfortunately, Michael was also the office IT guru, the person they relied on to resolve all problems with the lab's computer and networking issues. Maybe she ought to have someone from outside her area see if she could recover anything he might have sent to the Chinese. She'd just have to find a time when he was not there, which would be a problem, because Michael worked exceedingly long hours.

Dr. Langley had an hour to kill before she needed to get home and clean up. She planned to have dinner with Mark and his new girlfriend that evening, an event she had put off for weeks. As busy as she was, she always looked forward to her time with Mark. They could always find things to talk and laugh about. Their time together demanded little more than simple companionship. Besides, she was curious to see how Mark interacted with his new love interest.

Chaperoning two persons with Down syndrome on a date would be a new experience for her.

Ann flipped on her computer while musing about Mark. She absentmindedly searched 'Fertility clinic Minneapolis.' She found one in Uptown. It must have been the same one David went to. She wrote down the address and phone number on a slip of paper and put it in her purse. For the next forty minutes, she read and re-read the clinic's web page, and viewed their video on the procedure.

When finished, she sent off an email to David saying that he should consider working as a graduate teaching assistant in support of her new class on behavioral genetics next semester. She would be developing it as an online course and could use someone to help her with moderating real-time class discussions, along with related class administrative activities.

Ann met Mark and his new friend Lacy at the Olive Garden in Shakopee. They were already seated in a corner booth when she arrived. Mark rose and gave Ann his usual effusive hug, while Lacy sat composed and smiling in the booth. Ann slid into the booth next to her and exchanged greetings. Lacy placed her small right hand in Ann's, and her left on top and shook vigorously while looking over the top of her glasses with smiling eyes. "I'm so happy to meet you, Ann," Lacy said familiarly. "Mark has told me all about you. I think I like you very much."

"I've learned a bit about you from Mark. I'm want to learn more," Ann replied as she withdrew her hand from Lacy's clutch. Ann noticed that Mark seemed a bit nervous, which she did not expect from him. Whether it was being on an 'official' date, or eating in what he might have considered a fancy restaurant, Ann wasn't sure. She'd just have to go easy and make sure everyone was comfortable.

After the drinks and a basket of breadsticks arrived, Mark began to relax. Likely it had more to do with the fact that Lacy was so at ease with Ann. In fact, she was so comfortable that she did not hesitate to make a display of public affection for Mark.

Ann thought to herself that Mark and Lacy seemed like any other loving couple, perhaps even more like a long married couple than young people on a casual date. It seemed to her that there was no real reason they should not one day live together as husband and wife, at least with some level of support.

At 7:30 sharp, a caregiver from the group home arrived to pick up Mark and Lacy. Goodbyes were exchanged, and of course Mark gave Ann his customary parting hug. Lacy joined in, and the small gathering of friends headed their separate ways. The following day Ann received a thank-you email from Mark and Lacy. In simple language, both expressed an appreciation for the dinner and the good company. No doubt, Ann thought, someone in the group home had seen to this social gesture, but she knew that the sentiments expressed were Mark's and Lacy's. She was impressed that such care had been taken to teach them this formal social courtesy.

Taking advantage of Sunday morning solitude at the lab was Ann's normal practice. It enabled her to work productively. This Sunday Ann thought she might instead just have a quiet day at home. Her lesson plans were well laid out, and her lectures were now so familiar that she didn't need to run through them in her head. The final report of her research project still needed one last proofing before submission. Only a few graphs and charts remained. After a last review, it could be submitted to several leading journals for consideration.

Ann grabbed a cup of coffee and retired to the couch, a favorite place to just sit and think. As was her way, she started by mentally listing the issues she wanted to think through. She'd then sort them into categories. It's what a Ph.D. did, she thought. It's what any thinking person does. Name, categorize, and then sub-categorize as necessary. Ann would even debate with herself the order in which to ponder her issues. Of course, she'd have to decide on some criteria regarding the order in which to take them up. She had to laugh at herself that she didn't normally allow herself enough spontaneity to just let her mind wander. This time would be different, though. She'd just follow her thoughts wherever they led.

Ann remembered the emotional encounter she'd had with the department chair. Owing to its unpleasantness, she abruptly shifted her thoughts to the conversation she'd had with her brother about family genealogy. She turned that conversation around in her head for a time, and then realized she really was interested in knowing more about her lineage. If for no other reason, Andrew had piqued her curiosity about whether a distant relative had been committed to a mental institution. It didn't take long, however, for her thoughts to shift to David's sister's mental condition. Ann thought about how much she appreciated the beer and bull session with David. Although she'd had casual meetings with other advisees, she'd never had one in which a student was so open. Aside from his intelligence, she found his candor and his vulnerability charming. She knew such thoughts always, always had to be circumscribed with appropriate faculty/student boundaries, but she couldn't help it if the thought popped into her head. At least she could be free to think about his academic promise.

Ann got up to freshen her coffee and grab a muffin. Buttering it reminded her of her recent conversation with Lilith about having children. She knew she had failed to mask feelings of wistfulness at never having one of her own. Lilith had shown that she was feeling a bit sorry herself for her own life choices. Ann thought it was funny how food, like smell, could sometimes trigger memories and intense feelings.

Ann went into her spare room and grabbed the note she'd put in her purse. For a long time she thought about what it would be like to finally have a child using donor sperm. It wouldn't be their child too, but it'd be their seed … their genetic material … a part of who they both are, as well as those who preceded them. If she were to do something like that, given her age and the fact that she was unattached, she thought she would only want a healthy child. There was no way she could provide for herself and care for a child that had special needs. Even as she thought about it, she had to be honest with herself; she'd want a child she could pour herself into, one that had the capacity to grow physically and mentally, who could be somebody, someone who could get the most out of life. She'd want a child who could learn easily, and who could be fulfilled emotionally and intellectually. It would be good if her child was really smart. More than average, anyway. She wouldn't have to be a genius … no, nothing like that. But she'd have to value education, and be able to get the most out of it that they could handle … or could afford. Better than average looks would be nice, too … at least not repulsive. And since Ann cared so much about her health, she'd want her to be healthy, even athletic.

Running through the list in stream-of-consciousness fashion, Ann wasn't the least self-critical this time. She just allowed the thoughts to flow. Before long, she thought maybe she was being selfish, even a bit arrogant to consider the possibility. A baby, any baby, at least a healthy one, was to be cherished. Anyone would love and nurture life that they carried in their own body, especially if that child had some of their genes. Well, maybe not everyone, but she would. Anyway, she wasn't getting any younger. The odds were long. She might not even be able to conceive, or might not be able to carry a child to term. She had the resources. She could provide well enough for a child. Her career had reached a plateau at which she could at least pause. But then, maybe it was really at a dead end. Sure, it would mean giving up on some of her research, even withdrawing from her role in the consortium. She might have to curtail travel for a time. But other faculty raised children. They might not be setting the world on fire with their work, but at least they remained tenured. Maybe she could even give up her career as a professor — at least as a tenured professor in a large research university. She could always teach at a community college, or a smaller college that wouldn't expect her to carry on research.

What was she thinking? She had too much invested in a career to give it up now. She was more than halfway through a typical academic career as a professor and leading researcher, and had achieved some level of recognition in her field. It'd be nuts to walk away from it now. What would people think? She couldn't do that. Still there was something missing in her life. A lack of real fulfillment. Something that really mattered. Other than her brother and Liz, she had no meaningful relationships. What was more important? What woman really didn't want children? It was a biological pull. Hundreds of thousands of years of human evolution created the yearning. Summer would be the best time to have a child. She wouldn't have to miss much work. She'd have to coordinate with a number of people, especially inside her department. Her relationship with her friends would change. She was sure that Lilith would be supportive. So would her brother. She could manage. It wouldn't be easy, but she could manage. Arrgghh, this was never going to work!

She looked down at the phone number, steeled herself, and resolved to give the fertility clinic a call Monday morning. She could at least set up a free, no-obligation consultation.

That Sunday afternoon was unseasonably warm, a classic Indian summer day. Distracted by her mindless and unproductive morning inactivity, Ann finally showered, dressed, and took a slow late afternoon walk around the lakes. Returning home in waning light, she set about getting her mind right for Monday morning.

CHAPTER VIII

ABOUT ONE-HALF of Dr. Langley's Behavioral Genetics class was enrolled in a variety of graduate programs. All other students were upper division undergraduates. In addition to the mixed Tuesday and Thursday classes with undergraduates, the graduate students attended a Wednesday recitation session, where more advanced content was presented. Aside from additional reading assignments and more in-depth discussion, graduate students were required to write a 10,000-word paper summarizing research on a relevant topic. The mix of students on Tuesdays and Thursdays presented some difficulty for Dr. Langley inasmuch as the content on those days had to be challenging enough for her graduate students, but not beyond the ability of her undergraduates. Of greater significance was the need for her to adapt to the different levels of sophistication of these two groups. Yet, she was often surprised when her least well-prepared students, in other words the undergrads, offered the keenest insights.

A single hand stabbed the air. Dr. Langley peered over her glasses as students looked up from phones and laptops. An unintelligible murmur percolated through the room, then a hush. Ann did not need to hear what was being said; she knew what was coming, or at least thought she did.

"Yes, Rick," she said smiling. "A question?"

"No, ma'am … a thought." More murmurs … a few titters. It was axiomatic that the least able students too often managed to command the greatest attention. "I've been thinking about the eugenics movement," he began. "You once asked us to think about whether it had vanished from history. I don't think so."

"Please, Rick," Professor Langley encouraged sympathetically, "tell us more. How did you arrive at that conclusion?" She really did try to sound supportive.

"You take your average college student … no, let's say an Ivy Leaguer. I guess they're not so average. They come from money … they have smart and successful parents … parents who have good, high-paying jobs. Their kids are born smart, just like the parents. The kids are raised with all the advantages … good genes, good parenting, you know … parents who read to their kids … talk to them like adults … they go on educational trips. Later, the kids take a test and naturally score very high … most of them, anyway. They get into the best colleges. They meet and hang out with all the very best people … people just like them. They get good jobs, date and marry someone from one of those colleges. They move into the best neighborhoods … the ones with the best schools and the best students. Later, they have their own very smart kids, and the cycle repeats. Isn't this what the early eugenicists wanted? Haven't we partly organized society in a way that produces the result the eugenicists wanted? I guess the only problem is the rest of us."

"Well," Dr. Langley said, registering surprise, "you have been thinking, haven't you? What you're talking about is the effect of assortative mating. By the way, those neighborhoods you referred to have come to be known as super zip codes. What does anyone else think about Rick's observations?"

"Seems about right," someone piped up without raising his hand. "But, we didn't need eugenicists to manage this. We did it on our own."

"Makes me think the fears about genetically engineering intelligence are justified, even if we aren't there yet with the science," another student chimed in. "Who else but the very rich could afford it? And they're the ones who need it least."

"Even if what Rick says is true, it doesn't prove whether nature or nurture explains the concentration of higher abilities. Maybe the more successful families choose better schools for their kids, and offer a more nurturing environment. People have choices. Like he said, studies show they read to their children more, and things like that. Maybe if we just did a better job of raising and educating all children …" she trailed off without finishing her thought.

"That sounds great in theory, but we already spend a ton of money on schools and social services. We have for years. It sure hasn't helped test scores. The answer always seems to be to spend more money on schools and social welfare programs. Maybe genetic engineering IS a better way," another student responded. "What's so wrong about using science to improve society?"

"One of the main reasons we teach science is so that people are better equipped to make decisions about social policy. Let me remind you, our focus here is on science," Dr. Langley responded.

Dr. Langley did not deliver her prepared lecture. Rick's unexpected comments and more thoughtful classroom participation had enlivened a class of now eager students, some of whom were more accustomed to being fed knowledge than actively participating in its acquisition. It was refreshing for her to see how nearly all of the students had made some cogent observation, even in the absence of deep scientific grounding. However refreshing the class participation, it took less than fifteen minutes for the class to begin to sort itself into fiercely opposing camps: one taking the nature, and the other the nurture side of the debate.

The discussion sometimes got a bit overheated, but this time Dr. Langley managed to rein it in and to shift the focus back to science. To close, Ann said, "Look, this has been a good discussion, but I think we'll have to wrap it up here. You've shown a good grasp of the underlying impulses for the eugenics movement, and have made several keen observations. I have to say that we will not settle the question until, one: we answer the question scientifically, that is whether nature or nurture has the most influence; and two: whether we are morally equipped to respond, whichever the case. Even then there are sure to be skeptics.

"Now, I apologize for not delivering the lecture that I had intended for today. Your discussion was too important to neglect, but it does lead into what I originally intended to talk about. So next class, we'll be talking about several phenotypes besides phenylketonuria that, though they have a clear genetic basis, manifest differently depending on … I should say almost wholly on various environmental influences. Accordingly, please read the two articles that are noted in the syllabus. I'll see grad students tomorrow, as usual. My apologies for missing last Wednesday. Anyway, you already have your additional reading assignment."

◆

"First, let me say how pleased I was with your participation in yesterday's discussion. While it was totally unplanned, I think, at least I hope, that it might have done more to stir your curiosity than anything I had planned to present. As you could observe, sentiments about nature and nurture are no less intense than they were during the rise of the eugenics movement, and to a degree the questions then raised remain incompletely answered today. Let me just amplify a few points raised in our discussion.

"To his credit, Rick made a rather insightful observation about the factors that give rise to the question of whether nature or nurture has the predominant influence on social outcomes. At least in western civilization, the very structure of society has produced strong effects that together ensure perpetuation of class divisions. Yet the determinants of those outcomes, whether primarily genetic or environmental, cannot be teased from the class-sorting mechanisms we have in place. Rick might be right to suggest that, given the propensity of like breeding with like, some would argue that it is genetics that primarily gives rise to class distinctions. In that case, nurture may only provide a mechanism of support. On the other hand, it may instead be the more favorable rearing environment of the favored class that has the predominant influence. Yet, as we have demonstrated through twin and adoption studies, there most certainly is a significant heritability factor at work.

"As you have seen in your reading, Genome Wide Association Studies have now accounted for only about 15% of the heritability of general cognitive ability. Yet we know from twins studies that anywhere from 40-80% of IQ is heritable, depending on the age at which IQ is measured. Those studies show that IQ is more heritable among young adults than young children, suggesting that the effects of environment, of the rearing environment specifically, dissipate as one ages. We see the same pattern in other traits as well. Let me say, this is probably something you don't need to tell your parents, who may choose to think that they were mostly responsible for whatever positive qualities you possess.

"Now, the popular media, as with Solomon, likes to split the difference. Most report, if they report at all, that IQ is at best only 50% heritable. Unfortunately, that leaves the question unsettled, and is inconsistent with results from twins studies. Comparing GWAS data with twins data leaves a rather large gap in demonstrated heritability. Let me explain. With genetic testing, we have been able to calculate a composite polygenic score that gives us a singular measure of predicted IQ. In other words, we can sum the SNPs, the genetic variants that favor or detract from IQ into one score. That score explains only 15% of tested IQ differences. The remaining 85% we call 'missing heritability,' which might one day be found by running GWAS on larger samples.

"With more recent refinement to the polygenic scoring system itself, that is, by differentially weighting particular SNPs, we have seen minor improvement in our ability to predict IQ from genes. Let me boil this down for you, then. Today we have only slightly more predictive power in assessing a person's IQ from their genes as we have from knowing their family's socio-economic status. That being so, it would be wrong today to predetermine someone's path, say in early childhood, by looking only at their genes. Given the possibility of off-target edits using genetic engineering techniques like CRISPR, we would be deluded to think we could now alter someone's or an embryo's genes to yield a given IQ. Even with newer editing techniques like prime editing it seems implausible. The science just isn't there yet, and I emphasize YET. But, as we learn more about gene-to-gene interactions, understand the mechanism of gene regulatory elements, and especially epigenetics, our ability to determine IQ from our DNA will improve. That's also true for any other complex phenotype. Yet, one day....

"Before we go on, someone raised a point about social outcomes based on our genes. We talked about educational achievement and its relationship to our genome. In fact, several large GWAS studies have looked at that very question. Specifically, the number of years of schooling, like IQ, has a genetic component. That should not be surprising. People with higher measured general cognitive ability can handle more challenging courses, and pursue them to a greater degree, so that they are able to successfully complete more years of schooling. Polygenic scores for education achievement, EA for short, account for about 10% of the variability in educational attainment/years of schooling … which is slightly less than for IQ. Can anyone think why that would be?"

A young woman spoke first. "Assuming sample sizes are comparable, there are other factors that determine years of school. Ability to pay is one … maybe the most important. Also, parents' attitudes toward education could be a factor."

"I believe that's correct, Ms. Han," Professor Langley said. "Interestingly, the sample sizes for GWAS studies of educational achievement are larger than for IQ. So, one would expect that they would do a better job of accounting for variance. However, as you've observed, environmental factors must play a role.

"So, where do you think this leaves us as far as research in our field? Should we do more GWAS studies, say, with larger samples? Anyone."

"I suppose we could, maybe with whole gene sequencing, but it seems like we're seeing little return on our investment in these studies. We'd likely find a few more relevant SNPs, and maybe improve our predictive power a tiny bit, but we'd still be left with a rather large missing heritability. I suppose we first need to learn more about how our genome operates."

"Good," Langley responded. "Anyone else?"

"I was wondering if it would be worthwhile to look at those with exceptional ability. Maybe people with IQs over 150, or something like that. Perhaps we could look at those who score high on just one IQ subtest, but not others, or those with exceptional musical ability. Wouldn't that possibly reveal something new about the brain?"

"That's a good thought. In fact, some work has already been done in this area. Likely that's the kind of project my group could take on, assuming we can find funding. Anyone else have a thought?

"If there are no more comments, there's one topic we've not discussed," Langley added, "but we'll save it for a later day, though I should briefly touch on it here. We have accumulated considerable evidence that environmental effects do have an impact on gene expression, epigenetically, as we say. There is evidence those epigenetic effects can be transmitted through the germ line, and passed down to future generations. While most of this work has been done in mice and worms … don't laugh … we at least have evidence of the biological mechanisms that could make it possible in humans. The implications are significant, as you will see in your reading for Thursday's class, when we'll discuss the manner in which certain genetic predispositions are differentially influenced by the environment."

As students reflected on Dr. Langley's last statement, a student at the end of the long oval table rustled in his backpack and pulled out a newspaper clipping. "Before we go on, I wanted to share something with you," he said. "I found this article in an old copy of *Atlantic* at the barber's a couple of weeks ago. I wondered if it was something I should bring up in this class, but wasn't sure if it was appropriate. Given what we've been discussing … well, I thought I'd take a chance. Basically, it says that the rate of births of children born with Down syndrome has dropped by more than 90% in Iceland. In some places in Europe it approaches a 70% decline, and in the U.S. it's maybe 50% or so. In nearly all industrialized countries, Down's birth percentages continue to drop. A less risky and more accurate test for Down syndrome has resulted in more women being tested. Given the legal option and social acceptance of terminating pregnancy, more women are choosing not to give birth to these children. At first, I was very skeptical, so I did some of my own research. I looked up public health department records in several countries and found these claims were true. I can't help but think that the issues that what's-his-name -- Rick -- raised are valid. We haven't drifted far from the tenets of eugenics. This certainly raises questions about how we value human life, and which lives we value. If worth at the margins is to be determined by one's contributions to and dependence on society, that certainly demands answers about treatment of the elderly, and for that matter any number of other socially and economically dependent groups."

There was only the sound of shuffling feet. Someone finally dared to speak. "Look, Sean, I don't see it's anyone's business to decide whether a mother'll carry a child to term, particularly when that child ...." She didn't finish her thought. "It's just no one's business, and she has the right to choose."

Before Dr. Langley could intercede, another student chimed in. "That reminds me of something else, but it's still relevant to our topic. The other day, Rick referred to the way the most selective schools admit students with the highest test scores, presumably the highest IQs. That's not exactly how he put it, but that's the implication. Maybe you remember the famous court case a few years back. Harvard was accused of discriminating against Asians, who, it was argued, had better test scores, better grades, and were similarly accomplished in student activities. Because they were subjectively rated as 'quiet,' they were denied admission in favor of legacy students and the children of rich donors, among others. Rick's point about social engineering was largely valid, but at least in this case the less able seemed to be favored over the more able."

"And what conclusion do you draw from this?" Langley asked.

"I guess that sometimes social policy can work against a meritocracy, if you define meritocracy a certain way. Ironically, in the interest of racial diversity it worked against a particular minority group. In the words of Charles Davenport, that seems cacogenic."

By cultural convention, the class was unwilling to pursue a discussion on social policy regarding race, so the topic quickly shifted to the validity of test scores and IQ measurement. The exchange continued for another hour. Dr. Langley wove into the discussion a number of important studies and aspects of genetics and neuroscience. Each of the grad seminar days was more or less like that. They built on her classroom lectures, but more than exposing these students to a broader range of topics and supporting literature, they elicited in this group of advanced students the curiosity that might motivate them to more fully investigate the questions that were raised. Whatever their educational and vocational direction, Dr. Langley hoped her students could at least begin to distinguish between good science and the flawed popular science that occupied so much of the media.

When the period ended, Dr. Langley returned to her office. She saw that her brother Andrew had sent an email. He wondered if he might call her at home that evening with what he said were some startling revelations from the genealogist they'd hired. She wrote back that he could call at 8 P.M. She quickly skimmed through her other emails, and paused to carefully read one from her department head. Owing to necessary reworking of the space behind her office building, it said that she would have to forfeit her coveted parking space. Her department chair was thoughtful enough to point out that contract parking was available in the large ramp only two blocks away, a fact obvious to everyone who worked in the same building. Adding to his carefully crafted slight, he included the phone number of the parking office. Unable to contain her irritation, she shouted a vulgar invective that echoed down the hallway.

Gathering herself, she wrote back and thanked him for the heads up. Additionally, she requested a meeting to discuss the possibility of finding additional funding for one of her most promising students in a doctoral program under her sponsorship. Her email did not acknowledge her irritation over the parking issue. After finishing the note, she fished in her purse for the note she had stuffed there. She called the fertility clinic in Uptown and set up an appointment for the following week. As irritation begets action, it seemed as good a time as any to strike out anew. Her decision might have seemed impulsive to anyone else, but even as she dialed, Ann justified to herself that she had been weighing the prospects and consequences of such a move for a long time. It was, she thought, finally time to refresh her life.

Driving home, Ann again imagined what it would be like to carry a child that was partially someone else's. If it was David's, at least his seed, she knew she wouldn't be able to tell him, and he couldn't find out on his own. He was her student and advisee, and he was going to be working closely with her in the lab. Just thinking about him, though … she couldn't quite sort out her feelings. The gift of a child. How could you compartmentalize something like that? She knew she would have to maintain a professional boundary. To some, though, it could be scandalous. It could be career ending. She had to force herself to think about something else. Instead, she thought about what her child might be like. It would be the product of two relatively attractive and intelligent people. Still, nature makes mistakes. It could turn out for the worse. As a scientist she understood that. At least the fertility clinic had a way of mitigating risk. The way genetic testing was done, she would have the choice of discarding any cellular material that carried a risk of congenital disease and other things. She'd even be given the choice of gender. Maybe those wouldn't be such easy choices after all. Given her age, maybe she couldn't conceive. If she did, she might not be able to carry a child to term. She understood that her age was a factor in a successful outcome. Would that she hadn't waited so long. Surely fertility clinics had developed some ways to offer a greater chance of success. She'd known of women much older than herself who had been able to conceive. She wondered if the clinic would ever share the genetic testing details with her, or would even let her do the sequencing, if only for research purposes. Her mind raced. She did not dwell on any idea long enough to reach a rational conclusion. The prospect of it all was beginning to seem overwhelming. She wasn't sure she was ready for all that would follow. Better to set these thoughts aside and concentrate on something else. Oh god, it was garbage day. She realized that she'd forgotten to put it out. As she turned into the parking lot of her apartment, she thought about the timing; at least that was something she might be able to control. She possibly could time delivery for early in the summer. That would give her plenty of time to bond with the baby before starting a new school term. It was a good time to begin to pull back on her research, maybe even the consortium she had started. Someone else could handle it. With her research project nearly complete and no plan for what was next, it could work. Yes, now might be the best time, all things considered. Ann changed into her customary grubby eveningwear, grabbed a glass of wine, and settled into the couch to ruminate.

The phone rang. "Hello, Ann, it's Andrew. I can't wait to share some news. You're not going to believe it!" he exclaimed, laughing.

"Hold on a second, Andrew. I've just been sitting here like a lump. I completely lost track of time. I'm going to put the phone down and grab something real quick. I guess I forgot to eat. Hold on, I'll be right back to the phone." She got up and raced to the kitchen to grab a box of crackers. "Sorry," she said. "Go ahead, tell me. I can't wait."

"You sitting down? You got your wine? This is going to blow you away."

CHAPTER IX

"YOU REMEMBER ME TELLING YOU our great-grandfather had some connection to the Iowa Institution for Feebleminded Children? I think it was in Glenwood, Iowa. You fretted about whether he was an inmate? It turns out he worked there. Let me back up. The genealogist, Kristen Youngstad, found a reference to him in Newspapers.com. The article was written when our great-grandfather started working at the hospital. There's a picture of him and a Dr. Mogridge standing in front of the administration building. The article said that before the hospital hired him, Nigel Wellbourne worked at the Eugenics Record Office in Cold Spring Harbor, New York. It said he'd be teaching the hospital staff to administer IQ tests. Kristen's find turned out to be quite a breakthrough. As a result, she was able to trace census records and subsequently the naturalization records for Nigel. He was English. She said she'd send a copy. There's more. Kristen called the hospital. It's still there, but it's now used as a chemical dependency treatment center. She talked to the institution's director. Nearly all the old hospital records are stored in Des Moines. So Kristen called there to see if she could obtain a copy of any records on Wellbourne. She spoke with a person who works in the state mental health system. They were able to give her quite a good summary of the old state hospital network at the time Wellbourne was there. Turns out that Wellbourne's previous employer, the Eugenics Record Office, might also have records."

"This is stunning, Andrew. I almost can't believe what I'm hearing," Ann replied. "I'm familiar with the Eugenics Record Office. I touch on it in one of my courses. I just had no idea we had any family connection to it."

"That's not the half of it, Sis. You ready?" he responded. "Let me back up. Wellbourne was married when he came from England to the U.S. in 1912. His wife's name was Gemma. They had two children … one named Gracie, the older of the two. A son named Eugene was born in the U.S. the year they arrived. The son had Down syndrome — Mongolism they called it then. Wellbourne started work at the ERO right after their arrival. Wellbourne's wife dies in 1920. Nigel moves to Iowa, to the state hospital, with his two children the year after she died. Later, his son dies at the state hospital. Shortly after that, Wellbourne moves to California."

"My head's spinning," Ann said. "How do we know all this?"

"I'll send the details once Kristen has it all written up. I just thought you'd want the quick overview," Andrew said.

"I can't believe she was able to find all this out. Is there more?"

"Indeed. This gets interesting. So some time after the move to Iowa, when the daughter was still in high school, the son was admitted to the institution, presumably because of his Down's. There would have been no one to take care of him at home during the day — the daughter had started college, so the son lived at the institution.

Anyway, shortly after the son had been placed at the institution, he died in a tragic accident. Apparently, it was the result of a fall from a silo. At least, that's what the hospital record showed. He was buried on the hospital grounds. After his death, Nigel left the employ of the hospital and moved to California. He apparently left a forwarding address with the hospital."

"It's amazing that these records are still around," Ann said.

"It is, but ..." Andrew responded.

Cutting him off, Ann went on, "Let me see if I've got all this straight. Our great-grandfather, Englishman Nigel Wellbourne, married with two children, worked in eugenics, did IQ testing at a state hospital in Iowa, where his cognitively impaired son died, and he later moved to California. Is that about right?" Ann asked.

"That's the essence, yes," Andrew answered. "Kristen said she'd try to find out more about where he went and what he did in California. She didn't have a chance yet, but said it should be relatively easy to run this down. She said there'd be census data at least ... maybe more."

"What happened to his daughter?" Ann inquired. "That makes her our grandmother. It's funny mom never mentioned her. I guess I was either too young to remember, or ..." Ann replied.

"Kristen thinks she did not go to California. By then she was in college. She said that she'd try to track down information on her after she gets more on Nigel. In the meantime, she suggests that if we're interested, we can do some research on the Eugenics Record Office in Cold Spring Harbor. There's a genetic research institute there now. They may have old records."

"I'm not sure when I'd find the time, but it seems like it might answer some questions. How about you, Andrew? You think you might have time?" she asked.

"I doubt it. If I do, I'll let you know," he responded. "So, what else is going on in your life?"

"Oh, not much ... the usual. Before I forget, what's all this going to cost us?" Ann asked starting to register concern.

"Whatever it is, we'll split it, as we agreed. I think we're up to about $700 or $800. Is that okay with you? Since I gave her the okay, I'll be happy to …."

"That's all right, I guess," Ann said, starting to sound mildly peeved. "So how much more are you willing to spend on this?"

"I don't know. What say we cap it at another $500, and tell her no more travel for now. Sound okay with you?"

"That's fine, Andrew. But let's keep this under control … at least until we see her documentation and can figure out what questions we really want answers to," Ann finished. "Unless there's anything else, I've got to sign off. I've got some papers to grade before tomorrow. And, hey, Andrew. Thanks for taking the lead on all this. It's not the most important thing in my life right now, but as you know, I've gotten kind of curious about our family history."

"I have too, Sis. I'm glad to do it. Talk to you soon. Love you."

"Love you, too, Andrew. Talk soon. Bye."

It took a while for Ann to process all this new information. She had a thousand questions. What exactly was her great-grandfather's role at the Eugenics Records Office? How did he get involved in the field, which was in some respects the same as hers? And why had he moved his family to the U.S. in the first place? What had his life been like before he immigrated? As for Eugene and his sister, what were their stories? It was curious, too, that her great-grandfather worked in the field of eugenics while raising a son with Down syndrome. Was there some connection with his son's disability that motivated him to work in eugenics? It had always been a mystery that her mother never talked about her own mother. Then again, maybe she had, and Ann just didn't remember. Ann's parents had died in a tragic accident when she was only nine. Fortunately for both her and Andrew, the other grandparents were able to take them in and provide a good home. She thought there must have been some good breeding in there somewhere for she and Andrew to turn out so well. There really was something to this genetics, she thought, and chuckled audibly.

Checking her phone, she realized that her mind had wandered for nearly thirty minutes. She'd have to get on with grading papers in order to finish at a decent hour. Tomorrow was one of her busiest days, with three sections to teach.

◆

Ann had her introductory visit with the Center for Reproductive Services the following Monday. She was eager to start, having once made the decision to move forward. She was nothing if not decisive, at least after she'd thought things through thoroughly. Sitting alone in the small clinic office, listening to the sound of the receptionist typing, her excitement shifted to anxiety. She wrapped both hands tightly around the heavy ceramic coffee mug and forced herself to think that any reservations she might now be feeling would disappear once the process started. Despite having done a fair amount of reading online, Ann still had a thousand questions. She knew that so much of what she had read painted a rosy picture. The happy-family pictures, the pastel colors, and the reassuring words made it all seem as if zip, zap, you had a good chance of becoming pregnant, if not on your first attempt, then in subsequent attempts. It was not as if these clinics dared misrepresent anything. They were careful to state that *in vitro* fertilization or other reproductive alternatives did not always work. Fortunately, Ann could discern the difference between marketing hype and the way biology really worked.

At least Ann's age was below forty-three years, the cutoff for services, and she had no known medical conditions that might reduce her odds of success. She had read that by her forties, the number of a woman's harvestable eggs was down to something like 150,000-200,000 from around a million at birth. By the time she reached puberty, the average woman had lost half her eggs. Thereafter, she would lose around 1,000 per month. By the time a woman reaches forty, not only were there far fewer eggs, but the quality of those eggs diminishes as well. Through normal means, the average forty-year-old woman had only a five percent chance of conceiving without medical intervention; with intervention, only a 25% chance of success. By age 43, the chance of success with medical assistance dropped to 10%. Ann knew all this, and was still not dissuaded. Ann was familiar with large numbers and low probabilities. Long odds did not put her off.

She was jolted from her thoughts when an attractive woman in her late thirties appeared at the door to the inner offices. "Dr. Langley, I'm Dr. Christianson," she said as she shook Ann's hand. "Won't you come with me?" She gestured to the doorway and made small talk as they walked down a back hallway lined with pictures of happy families and smiling babies. Over the next hour, they became acquainted in what was an unexpectedly informal chat about their respective lives. Their talk, though it did not seem so at the time, followed a carefully constructed outline designed to assess the client's motives, commitment, and level of trust in what was inevitably to be a lengthy and emotionally fraught process. The intentionality of Dr. Christianson's words, masked by her warmth and openness, had the desired effect. Ann resolved that this clinic and this physician could be trusted with what maybe would be a nearly complete redirection of her life's arc, her refresh, as she understatedly called it.

Dr. Christianson needed to assess Ann's resolve and ability to see the process through to a mutually agreeable conclusion. She also had to be sure that her patient could endure the possible pain of a dream being dashed. Dr. Christianson outlined the major steps in the process only briefly. Those would be discussed in more detail during a formal consultation, which was the next step. After they concluded their time together, Ann received a three-ring binder that included more information on options for and methods of achieving pregnancy. Flipping through it quickly, Ann could see that it offered considerably more detail than what was provided on the web site. Dr. Christianson said that they would be discussing its contents thoroughly on their next visit, which was scheduled for the following week. On parting, Dr. Christianson assured Ann that she would be her primary clinician and guide in the process, and that together they would seek the most favorable outcome. Something in the way the doctor said it made Ann feel special, and that there was even a personal connection between them.

Ann was eager to get home and study the material. Over the next four hours, Ann read and re-read the material, committing nearly all of it to memory. She learned the whole process likely would cost $20,000 or more just to get to the point of pregnancy -- that is, if she were so fortunate. Some of the early expenses would be borne by insurance, but she would still incur major out-of-pocket expense. She quickly calculated that she could handle it, but it would put a dent in her retirement savings. She could at least pull this off without having to go into debt like so many families.

As to the timing, it looked like she might be able to start the process within the next few weeks. Getting through the physical exam, the shots, blood tests, sonograms, and more blood tests and sono-grams, could take another month, but that was optimistic. Three to four months was the average duration for the whole process, but that was just for one cycle. Some women needed three or more cycles to achieve pregnancy. She realized that a summer birth was unlikely, and that if she had any hope for it, she had to get started right away. Whatever the case, she knew she would need to adjust her workload for the ensuing year.

The rest of the manual touted the clinic's special expertise, especially its "unique, proven methods." She learned that they'd have to wait until the blastocyst stage, some five or six days after conception, before implantation, and that implanting one egg rather than multiples normally achieved greater success. But normal didn't apply to her since she was over forty. Although the likelihood of a successful delivery improved with only one egg, in most cases like hers, multiple eggs would likely have to be implanted. The prospect of raising twins or triplets wasn't something she had ever considered. Still, she resolved to press on.

Ann learned about cryopreservation, 'hatching,' ICSI, and other unfamiliar terms. She made notes to do some additional reading. There was a whole section on Pre-implantation Genetic Diagnosis (PGD) and Chromosomal Screening. Now she was getting into material with which she was familiar. From somewhere around the year 2000, reproductive clinics had been using various methods to assure that embryos would have the greatest chance of success in life. They could assure that fetuses would not carry any single gene mutations associated with congenital problems. Genetic conditions like beta thalassemia, Fragile X Syndrome, sickle cell anemia, multiple sclerosis, and Huntington's Disease could be prevented through screening. With chromosome analysis, various trisomy disorders like Down syndrome could also be prevented. Every step in the process seemed to have a line item cost, but each offered an advantage for the client as well as for the child.

With whole genome sequencing and the use of polygenic scoring, reproductive health clinics had recently began offering optional counseling on a wider range of other possible health-related conditions. Prospective parents could now get an assessment of probabilities that their child might have, or could develop medical conditions later in life. Whatever benefit this might offer, it also increased the complexity of decisions prospective parents might have to make in selecting embryos for implantation.

In vitro fertilization and other forms of reproductive assistance had become well-established, if not routine. Success rates could be reasonably well predicted, and the chance of congenital birth defects greatly reduced. With these new capabilities, reproductive science still required hard choices. In fact, they demanded that participants grapple with dilemmas challenging enough to induce moral vertigo. If embryos with probable future health problems were to be implanted, should the resulting children be counseled about known risk factors, and if so, when? At least one choice was simple, that of selecting the gender of a child.

When not teaching, these concerns took over Ann's thought life. And she hadn't even decided on a donor. Theoretically, David was an option, but one with what she thought had ethical and possibly career implications. Stressing, Ann knew she needed to vent. For that, Lilith was the best option. André worked nights except Sunday and Monday, so Friday might be her best option for another girls' night in. So it was arranged, and no costume would be required.

Ann outlined the complete *in vitro* process for Lilith. Throughout the explanation, Lilith was attentive and gave every sign of being supportive. She offered nothing in the way of her usual sarcasm, at least at the outset. Though for Ann, such commentary might have been a welcome way of defusing her anguish. When Ann finished outlining the process to Lilith, the conversation moved in a direction Ann did not anticipate.

"So, dear friend, all this screening and discarding as you referred to it, raises in my mind a number of questions. You ready?" Lilith began. If you know your embryo, or fetus or whatever had a chance, even a small chance of developing cystic fibrosis, a condition that often results in much suffering and premature death, would you terminate that life? Even if that child might turn out to be a lovable and handsome genius? And if you could know your little Langley had some probability of becoming a schizophrenic in his late teens, should that little Langley be terminated before he takes his first breath? You with me? Okay, then how about one that has an elevated probability of developing dementia, say at age 70 ... a keeper? That's maybe a long enough life to consider saving this one. But what if the onset was determined to be closer to forty, still a keeper? Heart failure, Crohn's disease, childhood cancer, or just plain fat and ugly? Some suffering for these, for sure. Maybe one whose IQ relegates him to a life bagging groceries? Where exactly would the line be drawn? And, who should have a say in all this? Just the birth mother? Look, this all gets at the question of what is a full, healthy and fulfilling life? Who gets to decide? Is it just a woman's right to choose? Does society have any say? And what exactly is suffering, or too much suffering? If we grant ourselves the right to terminate the life of an embryo or fetus because its prospects for a useful and satisfying life are dim, why wouldn't we do the same for the born, especially the elderly and infirm? If, as you've told me multiple times, we're all just a bunch of walking genetic typos, and humans are biologically designed to produce variation, why wouldn't we accept that variation is part of life? This all seems so consumerish. Like shopping for a child. Puts me in the mind of house hunting. This one's got a crack in the foundation. It won't do. That one's decorated wrong. Bad color. This one's got a furnace that is likely to go bad. Plumbing's a problem with this one. These just won't do. Why can't we just love and care for everyone, no matter -- or better said, because of their afflictions? But then, who am I to talk? Something to think about, you know what I'm saying?" Without waiting for a response, Liz got up to go to the bathroom. She planned it that way.

Ann was shaken. These thoughts had crossed her mind, if only briefly. The fact is, she pushed them away as fast as they arose. Hearing Lilith raise these questions now, though, somehow made them more real, not just hypothetical abstractions. Still, the probability of any of these problems was remote. Polygenic risk scores were not certainties. Women do *in vitro* all the time. We never hear about problems. Things normally work out, or they don't. Women have a healthy baby, or they're just not able to conceive. It works, or it doesn't. Besides, Ann wouldn't likely have to cull any embryos, except maybe for those the doctors say have no chance of surviving the womb. The choices would always be hers alone.

When Lilith plopped back down on the couch, she reached over and touched Ann's arm. She offered sincere assurance that she would see this through with Ann, whatever the outcome, and support her in every way. Ann was relieved, somewhat. She pushed Lil's little thought experiment out of her mind for now, to be taken up at a later time. It was the gift of a disciplined mind. Then Ann began to tell Lilith the part about 'cryobanking,' and the tone of the conversation changed.

Cryobanking was the lingo used to describe donor recruitment, specimen collection, donor selection, and the specimen retrieval process. Cryobanking — it all sounded so deceptively futuristic, so technical. Bolstered by wine and a predisposition to snark, they both knew that they could not get through an evening together without Lilith once again obliterating social convention, restraint be damned.

Ann grabbed her laptop and typed 'cryobanking' in the Google search line.

"What the hell's a cryobank?" Lilith asked with her familiar cynicism. "Is that like online banking, only with sperm? You make deposits and withdrawals? From a sperm kiosk? Where's the nearest kiosk? To make a deposit you take a picture and send it in via text message? Umm, excuse me, can you tell the balance in my account? Overdrawn! How can I be overdrawn? I deposit every day. Can I get a loan? I promise to pay it back. Hon, how exactly does any of this work?" Lilith went on like that for five minutes before Ann reached over and gently touched her arm.

"Are you through?" Ann asked. "If you'll sit back a second, I'll explain. Have another drink. What you're going to see on this site is nothing like the softly feminine, genteel way that the so-called reproductive services are presented. The reality is that all these sites, these sperm banks — I think we can call them what they are — look more like dating sites than something to do with reproductive medicine.

It works like this: The cryobanks pay men, mostly young men, mostly college men, to donate sperm. They donate once or twice a week for at least a year. The sperm is frozen. The client, me in this case, selects a donor and pays for a couple of vials or more. It's shipped to the clinic, though I could have it shipped to my home, and if I so chose, I could inseminate myself. In my case, the clinic would have to fertilize some of my eggs in a petri dish. Those eggs would have been frozen first— they say freezing helps. The eggs are then thawed, fertilized, and grown for a few days. DNA testing would be done on the embryos, one or more would be selected for implantation, and then we'd wait for nature to take over. That's it in a nutshell."

"That's all well and good, but how do you go about selecting a donor?" Lilith asked. The tone in her voice betrayed stunned incredulity at what she was hearing. "Do you get to meet the donor, or at least to get a good look at him? Dinner and drinks, at least? What about his siblings, or the in-laws ... especially the in-laws? Do you get a chance to check them out? Like I've always said, you don't just marry a husband; you marry his whole family, including his weird uncle and dumb brother."

"I think it'll be easier if I just show you. Promise you won't laugh. You promise?" Ann insisted.

"You know me, hon. Can't promise anything, but if I do, you're going to be laughing with me."

Ann pulled up a profile for one donor. Lilith scanned it and sank back into the couch, her eyes wide open. Gradually, a smile creased her face. Ann knew what was coming.

Lilith read the summary for donor #572173: "5' 10"; white; brown wavy hair; CMV, negative; safety engineer; blood type, A+; Ancestry, Italian/German. "Crikey! This does look like a damn dating site. At least you get to pick colors. What's CMV, by the way?" Lilith could not restrain herself.

"That's just a preliminary scan. You can dive deeper, refine your search. Look here," Ann said, showing her how to view more information for this donor, and how to refine a search for a donor.

"Um, you don't even get a chance to court first? Maybe check out his apartment?" Lilith was just getting started. "Safety engineer. What's this guy even look like? How's his build? And do you know if he's even got strong swimmers? By the way, what's a vial cost? Is it cheaper than dinner and drinks?"

"They're about $130. I'd need a couple, plus shipping. I get my choice of vial types for an added fee; you know, the good stuff costs more. You need to allow 14 days for delivery, Fed Ex."

"Sort of like ordering from GrubHub, eh?" Lilith commented. "I'll have two vials of the #572173 … and hold the lettuce!"

"Indeed," Ann replied. "Look here," as she scrolled through other donors. "Just to answer your other questions, their swimmers are evaluated first. If they can't perform, the donor is rejected. You can also see how many times a donor's been selected, and how many children they've produced."

"Now standing at stud, #572173, proven sire."

"Something like that," Ann said, beginning to laugh. "And no, there's no courtship, but I can see a picture. Take a look." Ann pulled up a childhood picture for #572173, the safety engineer. "Oh, and CMV stands for Cytomegalovirus. It is a test to see if they have or have had herpes, and if it's active. Negative is good."

"Oh, great! Think if I had the ability to check that for everyone before I dated them. Anyway, in this picture the guy looks like he's only seven or eight. Isn't there an age limit? Talk about robbing the cradle!" Lilith went on. She had already seen from the demographic summary that the donor was twenty-seven. "What does he look like now? The kid might have gotten ugly, you know."

"I have the option, at least in the case of some donors, of purchasing access to an adult picture. The kid picture is only to show what mine might look like in childhood. Some of the cryobanks now also offer short video vignettes. I can't show you one just now, as I haven't yet purchased an optional photo package. But I can show you a brief summary profile that's been written about this donor. You'll like this."

... is an adventurous gentleman. He likes to rock-climb and para-sail. He also enjoys fine wine, classical music, and a good laugh. ... is well established in his career, and is a careful planner for his future. Forthright, considerate and well-spoken, this polished individual knows what he wants, and is determined in its pursuit.

"Who writes this stuff, anyway?" Lilith asked, looking at Ann and then back to the computer. "This really is like a dating site. I had no idea this world existed."

"They all read pretty much like this. Look, Lilith, I agree with you. It's not the most dignified way to go about it, but maybe there is no dignified way. It's certainly not what I experienced at the clinic. Anyway, it's just something I'll have to deal with. I haven't picked a cryobank yet, let alone a donor. From what I've seen, they're pretty much all alike, though. You can select a donor from anywhere in the world, and pretty much see the same format. I expect I'll just pick some good old fashioned Midwest sperm, and proceed from there."

They shared a brief laugh before Lilith began another assault. "Moms always say they're happy if the baby is born healthy ... you know, has all its finger and toes. What bull! When given the choice, what they really want is looks, brains, and athleticism. Just look at these profiles. Tell me I'm wrong."

Ann sounded defensive in her reply. "Well, maybe that's all they've got to choose from if they have to go this route. When you have a baby naturally, you've got a bunch of things to consider. You know the father well, and meet his family. At least then you can make some assumptions about what your child might be like in a whole host of areas. Besides, mothers decide how they'll raise their child."

Lilith realized that she'd once again gone too far. She always did, but was not so insensitive that she could not see the effect. "Ann, hon, you've got to forgive me ... again. You know me too well to know that I can't resist popping bubbles. I don't mean anything by it, but you've got to agree, this whole thing is so weird. I know it's important to people ... to some desperate couple wanting a family, it ought to be approached with a bit more class. Look, I'm your friend. I support what you're doing. I swear on it. I promise to stand with you no matter what. It's the right thing for you ... and it's the right time. We both know it. Girl, I'm with you!" Lilith reached over and hugged her friend. Both had tears in their eyes. Ann turned away self-conscious at her display of emotion. She got up to grab another bottle of wine from the kitchen.

After Lilith left, Ann sat on the couch and scrolled through donor profiles. It seemed strange that each bio focused mostly on physical attributes. The limited and puffed-up personality profiles suggested minor variation from donor to donor. Even then, the narratives all seemed to have been written by the same person. They provided an image of men who were nearly completely lacking in flaws. Funny, she thought, women would have to base their selection on very little information, and of such dubious accuracy. Save for claims about years of education, which at best was only an approximation of intelligence, they had no way of knowing just what kind of smarts the donors possessed since IQ data, if collected, was not shared with. At least years of education was an indicator. It's why courting before procreation made sense. Wouldn't it have been better at least to collect family medical histories, plus do some genetic testing on the donor, rather than spend so much time on appearance? At least now it was feasible through whole gene sequencing to get some sense of the probability of future health problems. In the era of affordable sequencing, it was unimaginable that anyone should have to wait until the child was born to get an idea. For Ann to learn what she wanted to know, she would need to persuade the reproductive health clinic to give her access to the complete genetic sequence of her own embryos. Likely they would never agree, and she would thus not be able to apply her own polygenic scoring tool, whatever its limitations. It might prove useful in her research. Would that it were possible. No matter. Maybe David was still a viable choice as donor. He needn't know, and she knew she could keep it confidential. She'd just have to find a way to identify his profile in the sperm bank database. In the absence of donors' names, the brief narrative and college information might provide the necessary clues she'd need. If she was lucky, there would be a picture of him as an adult, maybe even a video. Then she could be certain of her selection.

CHAPTER X

ANN STARTED the *in vitro* process right before Thanksgiving. It took considerable resolve for her to endure the loathsome daily stomach injections and frequent feet-in-the-stirrups examinations. Each ten-day cycle of hormone shots would be accompanied at intervals by four or five inter-vaginal sonograms and blood tests. If after repeated attempts she was sufficiently fertile, vacuum egg harvesting could be scheduled.

In early February, three eggs were harvested. It was a disappointingly low number but not uncommon for a first go-around. After the three eggs were graded, two were discarded, being too incompletely developed for fertilization. The third egg was frozen in the hope that it might later be a viable candidate. Ann's reproductive health specialist explained that it's normally preferable to accumulate a total of fifteen eggs before fertilization can begin, though in the end, only one or two eggs would be selected for transfer. Accumulating the preferred total would normally take several cycles over more than a year. As consolation, the specialist said most clients usually only produce twelve eggs, and ended their egg production cycles around the one-year mark. Of the twelve harvested eggs, typically only eight are judged viable enough to freeze and later fertilize. Of those, customarily only six remain suitable for fertilization after thawing. From those six, only four reach the eight-cell embryo stage, and of those, only two or three reach the blastocyst stage. Reaching the blastocyst stage occurs five to six days after fertilization. That's when DNA testing can be done. Finally, one or two, or in rare cases three, embryos are selected for transfer. And all that was for a woman younger than Ann. Gauging these long odds from her depressingly low egg yield, Ann conceded to another cycle. That was the first of many difficult choices she would have to make.

Ann realized that a summer birth was impossible, and that her workload for the ensuing year would have to be curtailed. She reluctantly accepted the need to cut back on teaching, research and her outside work with the consortium. Consortium members would be disappointed, but could easily find a replacement. Still, the consortium was her baby. Quitting it would be hard. Her laboratory team would have to be cut. Some advisees would shift to other professors. A few post-docs, including Michael, would begin to search for new jobs. David's work assisting Michael would end, but she would ensure his retention. He'd be her teaching assistant, and she would still be his academic advisor. She felt impelled to keep him close. After all, he was a candidate to be her child's progenitor, and her benefactor.

Ann got a call from the clinic shortly after Easter. The eggs harvested at the end of her second cycle yielded only one that was suitable for possible fertilization. With only two viable eggs now, she would need to come in to discuss next steps. An appointment was set for first thing Monday morning. Deflated by the news, Ann knew that she would once again have to discuss whether to begin another cycle, or finally begin the fertilization process. The latter choice meant taking a very big risk that two eggs would be sufficient. Having already learned the odds of success, Ann agreed to another cycle. This time, she produced two more apparently good eggs. Following yet another consultation and approaching her 42nd birthday, she opted to end the process of egg harvesting. They could now at last plan for fertilization, and, providence willing, transfer. With only four good eggs and still long odds of success, she had simply reached the end of her endurance and finances.

As quickly as she had made the decision to proceed, she still needed to make her final donor selection. Choosing a medical school student for her donor, she placed an order for two vials, making the necessary allowance for delivery fifteen days hence. The timing was perfect. The product was to be delivered the week of her scheduled return from the China conference. Going through IVF, one could normally take the opportunity to observe fertilization. Ann chose not to participate in that interesting scientific, if not sexually sublime event. Instead, she would spend time recuperating from the trip.

The week before her departure, Ann received a call from her brother. He once again was the bearer of exciting news.

"Ann, I've got more news. Remember that Kristen was going to try to track down more information on our great-grandfather? Well, she's pulled another rabbit from the hat. Get this: After Wellbourne moved to California, he began teaching at a parochial school outside Los Angeles. He was a middle school remedial math teacher. I don't understand how Kristen's able to find all this stuff out. Well, anyway, it seems Wellbourne was highly revered by the students and the staff. They've been storing some of his papers all this time. Evidently, when he became very ill late in life and knew he was failing, he asked them to hold his papers until such time as any relative might come looking for information about him. I'm not sure why he did this, but the holy fathers remained faithful to his request. That seems kind of strange, but anyway, Kristen is FedExing what they have on him. I'll get you a copy as soon as it arrives. She also said something about a journal. Kristen wanted us both to know that when she talked with the head-master, a Father Benedict, he spoke of our great-grandfather so lovingly and with such familiarity that it was as if he had only recently passed away. Ann, he must have been a really good man to have had such an effect. I'm so excited about getting these things, I can't stand it."

"Gosh, Andrew. This is amazing. Despite all I've got going on right now, what with clinic visits and my trip, I guess I'm kind of excited, too. Well, maybe not as much as the chance to make a baby, but still …."

"How's all that going?" Andrew asked.

"I've had doubts off and on. It's sometimes a real drag. I'm determined, though. I just hope it'll all pay off. Now that I'm finally ready to be a mother, time is running out on me. It's taken a while to reach this point, but it's something I guess I've really always wanted it more than I ever knew. Having a baby the normal way would for sure be more pleasant, but at least this way I get a whole lot more choice in the matter."

"Choice?" Andrew asked.

"When I get back from China, we should have the DNA results. Then we can make a decision about which embryos are healthy enough to transfer. With the genetic screening, we decide which are most likely to result in a healthy baby. I also have to choose whether to take a chance on having twins or just a single birth. Given the odds, I'm, uh … they're probably going to have to transfer two embryos. Besides all that, I'll have the option of picking the gender."

Andrew was tempted to make a comment, but thought better of it. "Ann …." He started.

"What? You were going to say something snarky, weren't you?" She laughed as she said it. "It was going to be something like: 'It's just like picking a baby from Amazon,' wasn't it? Look, Andrew, if science gives us the ability to make these choices, why wouldn't we want to take advantage of it? Who wouldn't want the best chance of a healthy, happy child ... another Langley genius? Trust me, no one would go through all this and knowingly choose an unhealthy child."

"Sis, that wasn't what I was thinking. And I know you don't mean that about having a genius. What I was going to say was that I want what's best for you. I'm with you all the way, whatever you decide. You know that, don't you?" he said, now starting to sound fatherly. "Anyway, I'm just looking forward to being an uncle."

Ann didn't want to get into more discussion with Andrew about the DNA testing. Discussing testing that might find single gene mutations for any of thirty-some monogenic disorders, might have led to an argument or a hurtful comment.

Ann was not eager to make choices about all this information alone. She knew she could certainly lean heavily on her physician for guidance. The trouble was that she would not just be making a choice about which embryos would have a chance to become a whole person; she'd also have to make a choice about disposition of all the remaining embryos. Some might be frozen for her future use, donated to others, or even used for stem cell research. At least then some good might come of her decision. But that meant confining them to a sort of suspended animation, at least until she made a final decision on their disposition. Alternatively, she could choose to just have leftover embryos destroyed. The idea was not as repellent to her as it was to many. Science tended to engender callousness about such things. The choice wasn't a bright red line. It wasn't as bad as trading or selling them, as some had tried. Besides, thinking of them as undeveloped, frozen clumps of cells rather than future children — calling them by their scientific name — embryos — or better still, blastocysts — made thinking about her choices somewhat easier. Of course, she could always just avoid the decision for a time. It would be easier to refreeze the remnant embryos and not think about it. The downside was having to pay $1,000 per year for storage. The timing of the trip to China was good. At least for a time she could put off having to decide anything.

The week before her departure, a small box containing Nigel Wellbourne's personal effects was delivered. Ann set aside work in order to look through them. After a quick inventory, she sorted loose papers and general correspondence into categories for later review, and began to read through his journal. The first entry was dated 1912. Nigel, his wife, and child were shipboard, making their way from England to New York. It was interesting that he began journaling at this particular time. It certainly would have marked a particularly important turn in his life's path. By the time she finished reading, it was late, so Ann turned in for the night.

◆

"Andrew, did you read the diary? All of it? What did you make of Wellbourne's suspicion about Eugene's death? It sure seemed like he carried guilt at the loss of his son, and he fretted over its impact on Gracie."

"I'm not sure, Ann. He clearly suspected the farm manager's son had something to do with it, but as he noted, the doctor's autopsy report offered nothing conclusive. I guess we'll never know," Andrew sighed.

"What did you think about the part where he's reflecting on how much he'd neglected his family in pursuit of career ambition? It was kind of sad, don't you think? He was thinking about chucking it all, moving away to maybe teach or something. He said something to the effect of working at something life affirming instead of negating. He must have reached a major turning point in his life."

"Clearly he was under strain at the time. And why not? Eugenics had gone off the rails, and his wife and son were dead. Who wouldn't have felt the pressure?"

"Those letters, I mean, the ones to Gracie. Why do you suppose they were marked return to sender? And why did they seem to appeal for some sort of reconciliation? There had to be a breech in their relationship. Was Kristen able to find out anything about Gracie's life? By the way, what do you make of his relationship with Julia? She's more than casually mentioned a number of times. "

"As far as Gracie goes, Kristen hasn't done anything else. She said she's working on that now. At least from the letters we know that she was at Iowa State University. She may have been a student, or perhaps she could have worked there. Hopefully, we'll get something, but I agree with you that there was some kind of breech. As far as Julia, I didn't think much about it. I just assumed she worked with him. They must have been friendly, at least from the tone of his journal entries. That's not so surprising in that they were working at the Iowa state hospital together for a time. After she moved back east, she apparently kept him up to date on her life … like when she told him she had a baby. I guess we'll never know anything more about her, unless you're interested."

"I suppose you're right about Julia, but I'm not much interested in pursuing information about her. There's nothing unusual about them keeping in touch. More importantly, I hope we can find something out about Gracie's relationship with her father," Ann said. "I can't help but think there must have been some pain there. It's not like a daughter to not write her father back. Mom never told us anything about her parents … at least other than to say her father died when she was barely a year old. I guess they weren't married all that long before he died. I kind of remember she said something about her father being in cattle."

"I remembered hearing that he raised bulls, and sold the semen to breeders. Anyway, you ready for your trip?" Andrew asked.

"I am. My presentation's ready, and I'm set to leave in just a few days. I'll text you when I get there. I'll call as soon as I can when I get home. We'll start fertilizing my eggs shortly thereafter, and hopefully be able to schedule the transfer. I'll keep you posted. Anyway, can you let me know when you hear anything from Kristen? Even if I'm in China."

◆

Ann settled into business class. She stowed her briefcase carefully under the seat in front of her and leaned back. She hoped that being away from home might allow her to leave her cares behind, if only for a week. The trip was the culmination of a long research project and marked the beginning of a new life journey. Her private thoughts were interrupted when her seatmate slid into the seat beside her. He looked to be in his early forties, and unlike Ann, completely relaxed. He buckled up, smiled at her briefly, and turned to his opened his iPad. Ann stole a glance and recognized the passage. Not wanting to appear intrusive, or engage with him, Ann pulled her presentation out of her carry-on. She began to make a few marginal notations. Just into the first few pages, the steward came by and took drink orders, which gave her seatmate an opening to introduce himself.

"Hi, my name is Charles, Charles Goodwin. And you are…?" he said politely.

"Oh, sorry, I'm Ann. Please to meet you. I hope haven't taken too much room getting settled," she offered.

"No, no. You're all right. Where are you headed? I mean, besides China."

"Beijing," Ann responded. "To a conference. Just for a week." Ann wasn't sure how much information to exchange. She hoped he wasn't one of those passengers who wanted to talk the whole way. She needed to go over her presentation a few times, and then try to get some sleep. Her preference for avoiding conversation was apparent in her body language. As always, she was cautious about sharing her work with others outside her field.

Sensing reticence from her clipped response, Charles said, "Oh, please do forgive me. I promise to leave you to your work," he said and turned back to his Bible reading.

Before she could acknowledge his remark, the cabin steward delivered their drinks. The interruption gave her a moment to consider how far she wanted to go in responding. "Oh, that's okay. There's nothing secret about it. I'll be delivering a talk about my research in genetics and general cognitive ability. It's actually quite boring to most people," she said.

"Perhaps, but it actually sounds quite fascinating to me. Maybe later, if you're agreeable, you can tell me more … that is, if it's not a bother."

The steward came over the intercom and announced that the cabin doors would be closing, and that passengers should return the seats and tray tables to the upright and locked position. The interruption enabled Ann to avoid further conversation without appearing rude. She returned her computer to her bag and placed it under the seat. Exchanging smiles with Charles, she leaned back and closed her eyes while the plane taxied into position. She remained in that position until the plane reached cruising altitude. It wasn't that she intended to be uncivil. As she learned long ago, some people just wanted the minimum amount of information, and she knew she had a tendency to overwhelm people with her extended monologues. She didn't want to be impolite, but she did have to prepare, and she had a lot of other things to think about. She signaled her intentions once again by setting up her computer. Opening her presentation, and leaning into her computer, she began to think about the prospect of finally having a child.

CHAPTER XI

Dr. Langley presented her team's eight-year study of genetic linkage to cognitive ability. Although her research made no dramatic breakthrough, it reinforced polygenic scoring as a method for relating specific genes to trait variation, successfully added several dozen more genes to the list of those relating to intelligence, and provided the basis for several new avenues of research. Finally, it modestly improved the predictive power of polygenic scoring for intelligence.

In summarizing her work, Ann noted that GWAS research was finally beginning to close the heritability gap. At the conclusion of her presentation, Ann received a plaque from the newly named Behavioral Genetics Research Consortium. Feted later that evening at a celebratory dinner, she received many personal accolades as the organization's first honoree.

Running on adrenaline from the night before, Ann was picked up at exactly 8:00 A.M by her hosts from the Beijing Genomics Institute. After an overview of their current research activities, she would visit an experimental school built for the purpose of validating the Institute's research. She would return to the Institute later in the day to observe work with genetically engineered neurons.

The school visit was to give her a first-hand look at an extraordinary social experiment run by the Institute in what was said to be a public-private partnership, a term which, in the case of China, she judged to be oxymoronic. Students at the school were screened for admission before birth, their cognitive ability having been determined from polygenic scores and family genealogy. In other words, the Chinese were screening experimental subjects based on pre-implantation genetic diagnoses of embryos from selected families.

Ann was startled by the explanation of the experimental design, but held her emotions in check and said nothing. It was obvious that their approach relied in part on a polygenic scoring system, perhaps a predecessor to the one she had just summarized the day before. She couldn't help but think David was right that Michael had been sharing confidential communications with his Chinese counterparts. Hadn't she herself advocated for tailoring education based on a polygenic score? Her work was still too new to have been the basis for selecting these school children, but it certainly could be used going forward. However, polygenic scoring of IQ had been around for a while, only it was definitely less predictive than her current model. Now seeing her own policy ideas actually implemented elicited an unexpected level of ambivalence. What the Chinese had accomplished had seemed coldly calculating, and so ... so eugenical. American society hadn't even begun to adjust to these new advances, let alone grappled with the ethics of it. How was it that the Institute was able to fill a school with children whose DNA was sequenced before their birth? Where could they have gotten enough parents to go through *in vitro* fertilization?

Arriving at the school mid-morning, Ann was stunned by the scale and modernity of the complex. It had been designed by one of China's leading architects. The design was intended to facilitate learning, and also convey a sense of modernity blended with traditional Chinese values. With its central crystalline spire and colorfully embellished contemporary rooflines, the architect masterfully presented a blend of Han Chinese dynastic dominance with an expression of modernity that was representative of the new China. Arched classroom wings, structured to represent a bird taking flight, connected to a central administrative building. Together they surrounded an expansive, beautifully landscaped central courtyard. The extensive use of divided-light windows conveyed a sense of openness, and yet order and structure befitting the organization's mission.

Stepping out of the limousine, Ann's host, Doctor Liu, said, "Welcome to 'Reach Higher Academy.' It is inspirational, isn't it?"

Gazing up at the central spire and feigning awe to her hosts, Ann thought that fifty years after Nixon, the Chinese were still trying to convince a world that did not need convincing that they were part of the modern era. They were, in reality, trying to one-up the industrialized West.

"Yes," Ann replied. "It is stunning." Walking toward the entry, Liu explained that the school was built to accommodate small groupings of roughly similar ages. Ann noted that he did not say grade levels. Each group was to be made up of twenty-five students. Students in each were selected based on how well they matched each other in native cognitive ability. Teachers were chosen for their ability to meet the advanced educational needs of these carefully selected children.

Accompanied by a few Chinese watchers, Ann was directed to a large conference room in which high definition projection panels completely covered one wall. Doctor Liu projected a slide that explained the experimental design in greater detail. "Each wing is made up of students who from before birth were given a polygenic score ....

Ann began before he could finish his thought. "You said before birth ..." She, of course, understood from his statement that the students were the product of *in vitro* fertilization. She was hoping to elicit Liu's views on the broad social acceptance of their experiment, but before she could finish her question, Liu responded.

"That means," Dr. Liu responded, "of course, they are the product of *in vitro* fertilization."

"Of course," Ann said. She elected not to pursue her question thinking that it might better be discussed in private. The thought came to her as well that the Chinese government likely did not require broad social acceptance in the same way as the West.

"While one might have suspected our interest was to validate nature as overriding the importance of nurture, let me just say that years of research have already answered that question. We do not require more GWAS or twins studies for confirmation. We Chinese believe that it's past time to begin reforming education, but not in the way of the West. School choice, charter schools, open enrollment, equity and the like are mere shuffling. This will not enable the customization of curriculum to meet the needs of students of varying abilities. With compulsory education and mandated curricula, schools in the West must by necessity meet the broadest needs of diverse student populations. America and Europe, being egalitarian by culture, strive for inclusivity. That is not our aim.

"We openly acknowledge that students have varying natural abilities, so it is necessary to structure education to reflect those differences. Starting from pre-school, we can begin sorting children based on native ability. As we all know, general cognitive ability is about both intellectual capacity and a proclivity for learning. Rather than teach to the middle, or pace education for the lowest-level student, we must accelerate learning for our most capable students. I expect, Dr. Langley, you have had the experience in school of being bored by repetitive instruction. Perhaps you have been made to perform exercises past the point where mastery has been demonstrated. We must be honest with ourselves. Great achievements are made by a select few. They are the ones of singular ability, the outliers, if you will. The problem is that there are too few such outliers. Too many of our schools have been built for the great middle of the bell curve. This is not to say that those do not have worth. Advanced societies need technicians, clerical and blue-collar workers, though AI and robotics are rapidly changing all that. Rather than frustrate children in school by trying to pound ten pounds of salt into a five-pound bag — I think that is the American expression — it would be better to separate students unashamedly and tailor curriculum accordingly. This can begin at a very early age. That does not mean the others don't have value, or that they can't associate with those of higher abilities. We facilitate those exchanges through sports and other forms of social interchange. In fact, these things are cultivated in our school. That is in the interest of broader social stability.

"As to the means of sorting, you are obviously well acquainted with polygenic scoring. Beyond the improvements you have made with this technique, and beyond its particular application in very large extended family groups, something we in China are able to do, we have made other advances. These have come not from increasing sample size, sample sizes as you well know are already quite large, but with the application of mathematics and a greater understanding of genetic mechanisms. Most scoring systems derive from earlier methods that add or subtract values based on the presence or absence of certain SNPs, particularly in protein coding regions of the genome. Our approach utilizes complex mathematics, which even I can't explain. Moreover, we look at excitatory and inhibitory characteristics of non-coding regions of DNA as well, and even consider epigenetic mechanisms. Satisfied at the predictive validity of our methods, we can now rely almost completely on these methods. Let me wrap this up by saying that our ultimate aim is to unlock the potential and eventually increase the numbers of those with the highest abilities.

"Oh, and one more point. We will still be administering intelligence tests at intervals while we have these students. In that way, we can ensure that only the best are able to advance within our system." Liu's broad smile conveyed a sense of superiority.

"I couldn't help but notice that one building sits off by itself, just across the courtyard. What exactly is that building for? It looks like it might also be a classroom building."

"Very observant of you, Professor Langley," Liu responded. "That houses another group, the lower scoring students, the great middle, if you will. It is a group that we added after we began the initial experiment. Given the sometimes-challenging behaviors in this group, we find it best to keep their classes separate from the others. To that point, I must add that our groups do interact. In fact they do so quite well with each other. I should have also noted that we have an extensive control group made up of students in our regular school system. We will be tracking their performance over time as well. Is there anything else before we proceed?"

"Yes, there is." Ann said. "Do you intend to follow these students throughout their lives?"

"We do. We intend to track life outcomes -- chosen career, income, health, life expectancy, and achievements. We'll also be able to follow life outcomes for their descendants. To summarize, we are looking for a more efficient way of achieving the broadest impact on Chinese society."

"Remarkable." It was all Ann could think of to say, despite her inner turmoil.

"Ah, yes. I see our tea has arrived. This might be a good time to take a short break before we resume. We have only another fifteen minutes or so of background material before we visit some of our classrooms. I trust that will meet with your approval, Professor Langley."

Ann checked her phone during the break and saw that David had texted. "Please contact me as soon as you return home. I can't explain now."

Sensing David's concern about confidentiality, she responded with an emoji. After the break, the small entourage made the rounds of several classrooms. Ann could see that, at least to the extent these classes were representative, students were much more engaged in learning than was her observation of students in the U.S. The level of class participation stood out, particularly in the upper grades. These children also participated much more actively in group work compared to the average students. Ann also observed that much of the learning was structured around group problem solving. That was substantially different from her childhood experience. Even when she had seen American students work in groups, they tended to lean on one or two of the most capable students, whereas these students all seemed to be active participants. Even allowing for personality, the difference was striking, though it might also have related to an inherent cultural difference.

Following the tour, Ann and Liu, along with a single watcher, drove back to the Institute for a visit to several labs. Curiously, the other watchers vanished without explanation. In the first lab they visited, Ann observed genetic engineering work on neurons that were grown *ex vivo*. Here, they were evaluating the effect of gene manipulation on cell formation and function. In another lab, work was being done on small mammals. The goal was to assess the effect of gene manipulation on ability to perform simple learned behaviors. Following a lunch break, Ann toured the primate lab, where similar experiments were being conducted. Given the development of cognitive ability measurement instruments designed specifically for higher-order primates, it was possible to evaluate a select few genetic alterations on problem-solving ability. Liu stated that it was taking a while to perfect genetic alterations that did not result in off-target edits, because those sometimes led to various cancers. Dr. Liu assured Ann that they were making substantial progress in that regard. Ann was stunned at the Institute's progress, so much so that she determined that she would have to find a way to keep better informed about Chinese research programs. Following her visit to the two labs, the small group took another short comfort break. Over tea, Ann and Liu discussed future research plans for many of the other consortium partners.

Liu looked at his watch, and then spoke in a guarded tone. "Professor Langley, as our departure time is approaching ... um, we have maybe an hour remaining; I wonder if you would be interested in visiting just one more laboratory."

"Certainly," she said, anticipating that he had something important to share.

They took an elevator down two floors and walked to the end of a long hallway, where Liu gained access by having his face and thumbprint scanned. "As you can observe, this lab is doing work about which we are much more security conscious, for reasons you will soon see. I know it is not a surprise to you, but human embryonic research has sometimes been quite controversial ... though it is for the most part still tightly regulated. We are especially careful to comply with international standards for such work. As we discussed earlier, students in the Reach Higher Academy are the products of *in vitro* fertilization. They were genetically screened shortly after fertilization. In this laboratory, we are at the very early stages of looking at ways in which embryos might be ... um, enhanced genetically. As you well know, any such experimental embryos must be destroyed. We began embryonic research here as early as 2015. Those studies focused on certain gene therapies using non-viable embryos, and those edits resulted in many incomplete or off-target effects, including mosaicism. Then in 2018, using viable embryos, we achieved remarkable success. This time, we used a new base editing technique to make single letter corrections. Our research then was to correct Marfan's syndrome. I believe your President Lincoln was thought to have suffered from it. Of course, editing in the germ line remains controversial. So in this case the embryos were also discarded. Gene therapies for rare diseases like Marfan's have their place, but other lines of research have greater potential returns. Because of controversy, I trust you understand that our work here must be kept confidential. I'm certain that I can trust you in that.

"You well know that more complex genetic alterations can be fraught with problems, particularly with respect to off-target edits. Being pleiotropic, genes can affect more than one organ or cell complex, and can thus have unanticipated effects. Still, there is much we can learn through safe experimentation.

"You are no doubt familiar with one very narrowly constructed genome wide association study from several years back. The experimental group was comprised of those having measured IQs of at least 170, that is, five standard deviations above the mean. Though the experimental group was small, only 1,238 individuals, it was sufficiently large to identify one very promising gene for more study."

"I'm familiar with that study. Their work has not been replicated, though," Ann replied. "Finding subjects for such a study is like looking for a needle in a haystack, as we say. We must be careful not to draw too much from its conclusions."

"Though that study was of limited value, it did point the way to a new approach," Liu said as he smiled. "You would agree that replicating their work, at least that type of approach, is warranted. Given the importance of finding that enhanced expression of the gene ADAM 12 is associated with exceedingly high intelligence, further study could be useful. With a lower cutoff … say, around IQ 150, we might confirm the relevance of ADAM 12 to superior intelligence, and find other important genes. Regrettably, we know that ADAM 12 is also pleiotropic. It is implicated in a variety of maladies, such as certain cancers, asthma, liver fibrogenesis, and hypertension. Curiously, low levels of ADAM 12 are also associated with Down syndrome, but even that reinforces its role in cognitive ability."

"Yes, I am certainly aware of ADAM 12. But I know of no study that has been done using a lower cutoff score," Ann replied. She managed to hide her anger as she began to understand what he was subtly conveying.

"Let me be more direct. We have done such a study. As a result, we have begun testing the modification of ADAM 12, and a few other genes in viable human embryos," Liu responded. "The benefits are obvious. Much of our preliminary work was on primates. Perhaps I neglected to mention those were the gene modifications we used in the primate lab we visited earlier. We are looking at a similar enhancement in human embryos, but as I say, we must also assure that such edits have no adverse effects. If we can successfully prove that enhancing those genes is not harmful in the early stages of embryonic development, we can think about how that work might continue … again, of course, subject to international conventions."

Ann said nothing.

"I hope you will forgive me. I've gone on too long. Shall we go upstairs and check on our driver? It is time for your return to your convention, or the hotel, if you prefer."

◆

Ann sat through the last afternoon of the convention but only halfway listened to the presenters. She was too preoccupied with the pace at which the Chinese were moving. It was certainly faster that any other society was prepared for. As outlandish as the uninformed would find the Chinese experiment of culling embryos and educating the resulting children based on a polygenic scoring system for intelligence, she was more disturbed that she had not been told of the Institute's new genome study on high-IQ individuals. The consortium members had always agreed to share their research plans. Moreover, she had been contemplating the same kind of study at her own lab. The Chinese embryonic gene editing studies seemed to cross a bright red line. Genetic experimentation on embryos was to be strictly controlled, and the Chinese were now well beyond anything that was presently sanctioned by international consensus. No less troubling was her anxiety about the possible implications of David's cryptic email. Bowing to discretion, she would just have to wait for her return before investigating his issue.

◆

Ann received an email from her brother just as she boarded the plane for her return trip to the States. It read, "Sis, it seems like I'm always the bearer of some kind of shocking family news. Do not be troubled, but I must report a rather perplexing development. You may remember that I took an Ancestry DNA test last year. I've gotten a hit. Yesterday I received an inquiry from someone named Clifford Benson. He claims to be related. He apparently did a DNA test and found that we aren't just distantly related, but are closely related ... very closely related. He offered his phone number, and we agreed to talk. By the time you read this, I likely will have spoken to him. I'll give you an update on your return. Knowing you'll be jet-lagged, I will wait for you to call me. I have no idea what this is really about. I spoke with Kristen, and she had no clue either. Stay tuned."

On the return leg of her trip, at least between the many meals and snacks, Ann wore her blindfold. Uninterrupted, she would have more chance to think, her preferred activity. She didn't consider herself an introvert. Ann simply preferred a life of the mind over idle chatter. She was devoted to ideas, even those outside her own field of expertise. They energized her. They gave her new ways of seeing the world, and they gave her a way to test her own beliefs and prejudices. When she tired of exploring one idea, she took up another, and sometimes even tried to combine them into some new truth. To Ann, the truth had aspect. It was faceted, and was best viewed from many angles. People usually speak about truth as if it's a singular, one-dimensional thing. For Ann, any truth was also a series of connected things, each with its own faceted surfaces, and each truth was linked in a series of cause and effect relationships with other facts and other truths. She felt that this was the way a developed mind approaches matters. Ruminating in this way, Ann eventually succumbed to fatigue, and slept until the plane made its final approach into San Francisco.

◆

With a fifty minute layover, she found the nearest unused gate area in which to make a quick call to David. Her pulse quickened as she dialed. She expected his news to be upsetting. His voice calmed her, momentarily. "So, what is this important news that you couldn't share while I was in China?" she asked.

"I'm sorry for conveying the message as I did," he said. "It's just that, you know, you've always advised me to be discreet. I couldn't be certain our calls weren't monitored. Again, sorry if I alarmed you. To be candid, though, it does have a bearing on Michael."

"Please, David, I'm pressed for time. Can you just tell me what it is you learned?"

"Sorry. Anyway, Michael learned from one of his counterparts at the Beijing Institute that they are doing edits on human embryos. They've been doing this for a while now, after first testing a certain gene edit in primates. Evidently, earlier attempts failed, but using some new technique for editing … I think Michael said they're using a synthetic molecule … they don't have the same problem with off-target edits that CRISPR does. Michael's contact seemed to know you were visiting the embryonic research lab, and that's why he wanted you to be aware of what was really going on."

"How do you suppose he knew that, I wonder? That was my last stop, and my visit to that secure lab did not appear to have been widely known."

"Michael didn't say, but he also knew you were visiting the school. But that's not the real reason I wanted to talk with you. I know Michael had been sharing our research in detail with his counterparts in China. I overhead him having a conversation with someone in China. It was very late, but early in China, and they were speaking in English. I guess Michael thought no one else was in the lab at that hour."

"Sorry, David; I've got to find my gate. I can't talk any more just now. We can pick up on our conversation when I return, but please do not say anything about this to Michael. Between you and me, I've got to say that any trust I may have had in him was totally misplaced. You know what I'm saying?"

"I do," David replied.

"Anyway, I've got to be going. I'd like to meet with you when I get back to town. Give me a day or so to recuperate. I'll call you. Take care."

CHAPTER XII

First thing after Ann got home, she called Dr. Christianson, her reproductive specialist, to set up a consultation. The soonest she could get in was the following week. Frustrated at the delay in getting on with the next step at the clinic, she dashed off a text to David asking to meet early the following Monday. Next, Ann called her brother, still mildly irritated at what she presumed to be his relatively unimportant message.

"So, Andrew, what's this news you urgently had to share?"

"You sound irritated. You must be awfully tired."

"I'm okay, just pressed for time. Yes, and jet-lagged, too. Sorry."

"I'll be brief, but before I tell you what I was going to tell you while you were away, I've got something else. I just learned it yesterday. Kristen called and said she'd been scanning Newspapers.com. She was looking for anything else she could find on our great-grandfather. Searching for stories about the Iowa state hospital, she found another article. She said it was written not long after Nigel left the Iowa state hospital. It was about a teenager named Horn. Evidently, he was convicted of physically abusing inmates at the hospital. He'd been working for his father on the hospital's farm. Given the notes about Eugene's cause of death ... well, you've got to wonder."

"I've always wondered about that, but I'm not sure how we'll ever really know. It looked suspicious, but Eugene's death could have been just an accident. This Horn kid's conviction may be totally unrelated — anyway, what a twisted little bastard that kid must have been. What else?" Ann asked.

"Okay, so Kristen also found that Gracie was a professor of animal science at Iowa State University in Ames. Her focus was on large animal breeding. Besides that, you remember me telling you that I did an Ancestry DNA test two years ago. As part of it, I put our family tree on the Ancestry site, at least what we know of it, and made it available for others to see. That way, if someone's DNA matches mine, they could look at our tree and contact me. Like I told you, I got an email last week from someone related to Julia Benson. You remember that name?" Andrew said.

"No, not really," Ann replied.

"She was the one that worked with Nigel at the state hospital. She's mentioned multiple times in his journal."

"I kind of remember. You've got to excuse me; I'm still beat from the trip. Now that you mention it, I do remember her name was mentioned in his diary. Weren't there some letters from her, too? So who was the person who contacted you, and how were they related to Julia?"

"Ann, this gets more interesting. The guy's name is Clifford Benson, Dr. Clifford Benson. He just retired from a career in family medicine, and picked up on an earlier interest in genealogy. We talked for over two hours after we made contact. His father, Kent Benson, was born in 1927 and is the son of Julia Benson. Julia's the same one that worked with our great-grandfather. Dr. Benson has all of the records to back this up. He said he'd share them with us. Not only did Julia work with our great-grandfather, she lived with him in Iowa right after Nigel's wife died. Kristen was able to verify that from the hospital's records."

"Jeez, Andrew, my head's about to explode!" Ann said. "Are you sure about all this?"

"I'm certain of it. Dr. Benson even called Ancestry and got to speak with someone that knows about DNA. According to that person, the SNPs don't lie. We're genetically very close ... he said something about centimorgans, whatever they are. I suppose you already know about that. I don't know what to call Dr. Benson ... step-brother, half-brother twice removed, whatever. We've got to meet him. I'm sure he'd be interested in seeing Wellbourne's papers, and especially the diary and the letters he kept from Julia."

"Good idea, but can we put it off until sometime after I figure out where I am with the clinic? Let's just make sure this isn't some kind of scam. Is there any way we can do that?" Ann insisted.

"I think so. I talked to Kristen, and she said she'd run down information on Dr. Benson, but I don't think the DNA tests lie; do you? Meanwhile, Benson's agreed to share what he has on his family tree. I'll get back to you when I hear something. Anyway, I've got to go. I've got a patient in the lobby. Talk soon. Let me know where you're after your doctor visit, will you?"

◆

Dr. Christianson greeted Ann warmly. For the first few minutes they talked about Ann's conference in China. Naturally, Ann left out the parts about the Beijing Genomics Institute. After small talk, the doctor redirected the conversation.

"Okay, Dr. Langley, I thought we'd take a few minutes to review the next steps," the doctor said. For the first time since they began meeting, the doctor sounded less like an empathetic counselor and guide, and more like an emotionally detached physician. "I know this process feels lengthy. It is, at least to get to the point where we are now. For some it's two full years. Many women have more options at this point, but then again they have time on their side. Some are able to delay the next steps, for whatever reason. We do not have the same luxury. So, you ready?" she asked.

The clinic staff had always appreciated working with Dr. Langley. She was more mature than most clients, and had the background that assured them she completely understood everything she was told. Though as motivated to conceive as their other clients, the imperative for Ann did not appear as acutely felt as it was for most of the young couples they treated. Ann was a bit more prepared to be disappointed. Though hopeful, she would not be crushed by failure. Ann had the capacity to make the choices that were for so many almost completely overwhelming.

The doctor continued. "Your donor's vial arrived just two days ago in perfect condition. The timing was excellent. We've already conducted tests, and it looks like you've made an excellent selection. Motility and volume are both quite good. With your permission, we can begin to thaw the eggs. I recommend that we thaw and fertilize all four, and then conduct DNA testing on each. I have to caution you, sometimes some of the eggs do not make it through the thawing process. I recommend we fertilize all that make it through thawing. Do you agree?"

"I do," said Ann.

"From four fertilized eggs, assuming all remain viable for implantation, you would select two based on any gender preference and DNA results. In selecting embryos for implantation, then, it is always advisable to first consider the prospect of successful implantation and conception, beyond other considerations, that is, unless we find major DNA problems. Specifically, chances of successful delivery should normally override other considerations, again, unless we find major problems. Okay so far?"

"I am," Ann replied.

"Now, let's assume we're at the stage where fertilization is completed. As to any remaining embryos not selected for implantation, you can choose to donate them or have them discarded. As far as donation, you could make them available to another prospective mother, or for research. That again is up to you. We do not need to make that decision now. I simply wanted to remind you of your choices so you have time to think about it. If you choose to freeze embryos, or for that matter refreeze any eggs before fertilization commences, they can be cryobanked indefinitely at a small monthly fee. Many younger women come to us just because they want to retain eggs or embryos for their future use. Given your age, that is probably not an option. Anytime after freezing eggs or embryos, you can choose to discard or donate them. Okay?

"So, then, once we are prepared to begin fertilization, you may elect to observe that step if you wish. Some women find that an almost sacred experience."

"I don't think it necessary."

"We'll let you know when we begin the process should you reconsider," the doctor responded. "When fertilization is completed, we wait. This is always hard, the waiting, I mean, but it is necessary for embryos to develop five or six days before we extract DNA. When done, it will take another week or so for us to get the results. Normally from the date we complete fertilization, we can set a set an appointment to discuss DNA results, and also an approximate date for implantation.

I'm sure you'll want all this to plan your work schedule. Before you leave today, we'll print out a tentative time line for you."

"Doctor, I have an unusual request. As the DNA is mine, and as DNA research is my field, I'd like to request that I be given a complete run of the DNA sequences for any viable embryos we finally consider for implantation. It's not that I don't trust the DNA testing lab's analysis; I do. It's just that, since I can understand the results, analyzing them might over time give me empirical insights on my own research. Again, let me assure you that I have the ability to read and understand the results. In fact, we do the very same kind of sequencing in my lab. I'm prepared to sign a release if that is required."

"Frankly, Dr. Langley, we've never had such a request. It is out of the ordinary, but I promise to look into it. Of course I'd have to check with our DNA sequencing provider, and if our corporate lawyers see no problem, I think at least we can consider it in your case. I myself do not have a problem," Dr. Christianson said. "Choosing embryos is the most important decision you will make in this whole process. We want our patients to make the most informed choice possible."

"I understand, and completely agree," Professor Langley replied.

"One more important point. We find that once selected for transfer, freezing the new embryos give us the best chance of success. That process normally takes another week. When we get to the point of selecting a transfer date, you'll begin another series of shots to prepare your uterus. This improves chances of implantation. Thereafter, we'll monitor your progress through periodic ultrasounds. Agreed?"

"Agreed," replied Ann.

"So then, Dr. Langley, we can move to the next step tomorrow. It does not take long to thaw eggs or sperm. By afternoon you should have conceived," the doctor said enthusiastically. "We'll call you when we have completed this step."

"I'm eager to begin," Ann said, beaming at the prospect.

The next day, Ann received a call that conception had been completed on three eggs. The fourth egg unfortunately could not be fertilized, because it was judged be of poor quality after thawing. Ann elected to have it discarded. Mildly disheartened by learning of this setback, Ann was somewhat buoyed by the clinic's reassurance the remaining embryos were growing nicely.

By the fifth day, Ann received further encouragement that the remaining three embryos were sufficiently developed to begin DNA extraction. Ann was scheduled to come into the clinic to talk about the DNA results in seven days.

◆

Ann met David in her office at 6 A.M. as scheduled. "I am reluctant to have to tell you what I've concluded, David. I know how closely Michael has worked with our partners. It shames me to say that what Michael and our partners have been doing is not what we once thought. The Beijing Institute has been running a massive social experiment on school children. Those children were selected based on polygenic scores. Though I can't prove it, it looks to me like they have used an older version of a polygenic intelligence scoring system to select the best embryos. Those embryos became the students in their experimental school. That part is true. That's exactly what they told me. But there's more. I believe they're now also testing genetic alterations in one small group of embryos. Those embryos are being tested and sorted using our newest polygenic scoring system, even before our research is published. The best of those embryos are being altered by making changes to at least one gene, maybe more. I suspect those altered embryos will be implanted rather than discarded. I just can't prove any of it."

David listened in disbelief.

"You can't say anything about this to anyone. This is extremely serious. I'm sure you understand. Not anyone."

David said, "I didn't tell Micael I was contacting you in China, obviously, and I definitely wouldn't say anything about what you're telling me now. I get it. But what are you going to do?"

"Look," Ann continued, "We can't be certain if any of this is true, other than the school. But I know what I saw. I saw the lab where they were doing work with CRISPR. They were editing human embryos for expression of ADAM 12. They assured me those embryos were being destroyed. I also know they've begun testing edits on this gene in primates. They laid out their whole genetic testing paradigm for me. It was almost more than I could do to keep for calling them out to their faces. But unless Liu confirms any of this, I'm not sure what can be done. "Liu … I don't know," she trailed off.

"What?" David asked.

"He was just so condescending. I began to realize it on my tour. I guess I saw a different side of him. We'd never actually been together all that much, other than seeing each other briefly at conferences. We've mostly talked by Skype. He's always seemed to be quite open, trustworthy. This time … well, he was different."

Maybe it was all a misunderstanding, Ann thought. She couldn't prove Michael had done anything wrong. Michael would be leaving her lab anyway, now that her work was ready to submit to several journals. Perhaps it was best to leave well enough alone.

Normally, Ann could think with detachment. This time, though, she was forced to reckon with the real possibility that a red line had been crossed. However serious, a false accusation would do irreparable harm, possibly even retard research worldwide. Then, too, there was the possibility of legal action against the university. The Chinese would surely deny everything, and it would be nearly impossible to conduct an investigation there. If she were proven wrong, it could be career ending for her. It might be even if her allegations were found to be true.

Ann once again swore David to secrecy and advised him that she would have to think about next steps before taking any action. She concluded their meeting by saying, "This thing has to be addressed ... somehow, some way. We can't let it go unregulated."

For Ann, this whole thing had become infuriating. It was a betrayal. More to the point, the Chinese were possibly loosing something that might never be stopped. Everyone in the scientific community had agreed not to introduce genetic alterations in the germ line. Safety of the subjects, and even safety of future generations, was at stake. What the Chinese appeared to be doing went beyond alleviating suffering. It wasn't like taking a risk to correct a gene for hemophilia. The Chinese were seeking advantage, dominance even. There was no other explanation. If the Chinese could get away with it, how long would it be before other nations began to muck around with the germ line? Then again, Ann wondered, maybe they had already started. At that point, Ann had to pull herself up short. She realized that her emotions were overtaking reason. She would just need to step back, more calmly evaluate the situation, and consider alternatives.

◆

"Lil," Ann began. Her voice betrayed a sense of urgency. "What are you doing?"

"You mean right now? Something wrong? You sound upset."

"You free to come over?" Ann responded.

"Now?" Lilith paused. "Sure, honey, but what is it."

"I'll tell you about it when you get here," Ann said.

"I'll be right there, but I can't stay too long. I've got to pick André up when he gets off work. His car broke down. Should I bring wine?"

"Not this evening. It's a bourbon kind of night," Ann laughed.

"Sounds serious; I'll be right there, hon."

Feeling overwhelmed, Ann needed a friend. Out of anyone she might have chosen to share her concerns with, Lilith was the only one who could be trusted to keep things confidential. While sometimes blunt, her criticism would surely offer a different perspective.

Lilith poured them each a drink, then they sat at the dining room table where Ann laid out the whole story of her visit to the Beijing Institute. When she finished, she asked Lilith what she ought to do with the information.

Lilith took a deep breath. "What is it with people? We always think we can do better than Mother Nature. I'm not talking about healing the sick, ending affliction, whatever that is. I'm talking about the stuff we do to make it better only for our kind, our people … those on our team. Give ourselves a leg up, you know what I mean? There's nothing organic, nothing natural about it. Weren't we engineered to produce variation? Isn't that what billions of years of evolution have done? Did evolution get it wrong? Isn't that they way our genes work? Haven't you told me we're all just a walking mess of genetic typos? Look, we've survived haven't we? Besides, aren't we called to care for one another, especially the sick and the disabled, especially the disabled, whatever disabled means? Is the goal to prevent their very existence? And as far as intelligence, we're surrounded by smart people, book smart and life dumb people. Don't get me started. Look, this stuff you're talking about won't end well. This kind of thing never does. I don't know if any of the allegations are true. I have no way of knowing, but I can believe it. It's not just a Chinese problem, though. Science … technology … they've given us so many choices. Maybe too many. It's the myth of progress. We think that we're somehow smarter now … wiser. Progressives love all this stuff. Somehow they think they can engineer things, and poof, make it all better. The real problem is, and it's not a national problem or an institutional problem, it's a human problem, a heart problem. The real problem is that despite all this technical progress, there's been little moral progress, maybe none. I don't know if we're equipped to handle the choices we assume for ourselves. So dear friend, I'm not sure that's what you wanted to hear from me. But that's about all I can think of for the moment." Lilith took a swig from her glass and declared, "I've got to use the little girls' room. Geez girl, you've got me wound up now," she said as she walked away.

Ann was dumbstruck. To her, Lilith seemed to miss the point about whether to report what she'd heard about the Chinese implanting altered human embryos. When she returned, Lilith took a sip of bourbon and began, "So, other than that, how's your test tube baby project coming?"

Ann wasn't put off by Lilith's wisecrack; it was to be expected. "I go in tomorrow to discuss the DNA results. I've got three viable chances to make a baby. I have to choose two, and then schedule the transfer. There's some shots, then the deed will be done."

"Can I ask? Was it was good for you?" Lilith joked.

"Lil, you're such a pig," Ann responded.

"So anyway, two out of three, is it? Those seem like good odds. What do you think, girls or boys? One of each? You do have a choice, right?"

"I'm told I do.," Ann said.

"You got a preference?" asked Lilith.

"Not really. It comes down to which ones give the best chance of delivery ... that is, if the DNA says they're healthy."

Lilith looked at her watch, grabbed a cracker and cheese, stood and announced she had to get going. Ann was left wondering just what it was that Lilith had offered in the way of considered wisdom. She poured herself another small drink, and pondered her options.

◆

Leaning across the small table and touching Ann's hand, Dr. Christianson said, "Well, Dr. Langley, This has been quite a journey hasn't it? We now come to the point where it's time to make some choices. Before we get into the DNA results, let me just give you an opportunity to ask any questions you have, or comment any way you wish."

"I don't really have questions at this point, but I do wish to say that it's been a pleasure working with everyone here, and you're right, it's been a real trip! I guess it's time to take the last few steps."

"Let me show you what we've found from our testing," the doctor said as she projected the image of Ann's results on the wall monitor. "As shown in these images, we found nothing untoward in any of the three remaining embryos. Each developed quite normally up to the time of testing and subsequent freezing. We have every reason to believe that each will come out of freezing just fine, though of course we cannot guarantee that. Now, before I show you the next slide, let me caution you that the results from the chromosomal tests represent only a partial analysis. Other DNA findings, which we will discuss in a minute, may dictate the choices you ultimately make. Let me be direct; we have found something troubling in the chromosomal tests. I will not beat around the bush," she said as she reached for the keyboard. Before striking the key, she continued, "One of your embryos, a male, has a third chromosome 21 … trisomy 21 … which produces Down syndrome." She brought up the slide of the test result to confirm what she had just said.

Ann stared at the slide, pursed her lips and said nothing. The result wasn't altogether unexpected given her age. Ann turned back to the doctor and asked, "And the other two?"

"Each looks to have nothing that is particularly noteworthy. One shows a slightly elevated probability for glaucoma, and the last one seems to show a somewhat elevated risk for high cholesterol. Of course, these results by no means suggest certainty, and in any event, both conditions can be managed medically should they appear later in life."

"And the gender, for the other two?" Ann asked.

"One is female, and the other is male," the doctor said, smiling. "Many parents find this an ideal outcome," she added. "You may make a selection now, if you wish; or, based on our previous agreement, we can provide you the complete genome sequence of these two embyros for your own evaluation. After that, you can make your final decision. All three embryos will remain frozen until then … that is, unless you elect to discard or donate any at this time. Given the options remaining, at least based on what we know today, your choices are now a bit more limited than we had hoped. I'm sorry. In any event, the choice on how to proceed is yours completely."

"I completely understand. I'm anxious to get on with this, so I shouldn't need more than a week to decide," Ann asserted.

"You may advise us of your decision by phone or text. Just refer to the code numbers on the respective gene sequencing reports. We would schedule you for shots, and at the same time schedule the transfer procedure. Follow-up visits will be scheduled thereafter. So, if there are no other questions, we will await your decision. You can pick up the thumb drives of DNA results at the front desk on your way out."

Ann and the doctor stood and embraced. Ann left with the DNA results and headed directly for her lab.

EPILOGUE

Aɴɴ ᴅɪᴅ ɴᴏᴛ ᴄᴏᴍᴘʟᴇᴛᴇ the polygenic scoring of the two unimpaired embryos after all. After much introspection, she concluded that the scoring would not make any difference to her decision. She could make reasonable enough assumptions about the intellectual capacity of her children, given what she knew about the donor. More importantly, she had come to accept that she could love and nurture her own children whatever their inherent capabilities. Ann notified the clinic of her decision to proceed with embryo transfer of the two unimpaired embryos.

She elected to keep the embryo with Trisomy 21 frozen indefinitely. She had no expectation of ever using or donating it for research. Mostly, she was emotionally unprepared to destroy it. Transfer of her two good embryos was scheduled for the following week. The morning of the scheduled transfer, the clinic brought her disturbing news. During retrieval of the small rack of containers holding Ann's embryos, the clinic noticed that two of the three embryos had somehow catastrophically failed during the last freezing cycle. Those would have to be discarded. Before Dr. Christianson could tell her which embryo remained viable, Ann abruptly cut her off, insisting she was prepared for implantation whichever embryo remained. She simply did not wish to know which one remained viable.

"Ann," Dr. Christianson began, "I'm afraid our options …."

"I understand."

"What would you like to do?" the doctor asked.

Ann did not answer at first.

"We can begin another cycle if you like," Dr. Christianson offered.

"We could," Ann said, "but I'm not willing to go through all that again."

"You don't have to decide today. You can take some time to think about it if you'd like. We can freeze the last egg until you're ready," the doctor suggested.

"Look, I've had plenty of time to think about all this — even what I'd do if something went wrong. I've already considered the possibilities. As a matter of fact, I've had plenty of time to think about a lot of things. I'm prepared to proceed with implantation."

"Are you certain, Ann? Are you sure you don't want to know about the last embryo? You do understand what that might mean, don't you? You have a choice, you know."

"Choice ... right. I appreciate that, but it was enough that I could choose to be a mother."

"Ann, I'm sorry. I know this is hard. This outcome is not what we hoped for. We always expect to be able to give our patients more choice. And I'm sorry if I sounded patronizing ... for pressing the point. I just have an obligation to make you aware of the possibilities."

"I understand, and I'm not offended," Ann replied.

"So then, we will proceed in accordance with your wishes. Just know we'll do everything in our power to see that you deliver a healthy baby," Dr. Christianson reassured.

◆

Following announcement of her pregnancy, Ann resigned from her role as director of the consortium, and reduced her teaching load. Any plans to secure grant money to begin new research were put on hold. Putting off research to raise a child, Ann knew, would make it more difficult to later secure funding for any major new research project. All of it would change the trajectory of her career. Ann would remain a graduate advisor and take only one advisee for the time being. David would continue to work closely with Dr. Langley as her advisee and graduate teaching assistant.

Three months after delivery of a son, Ann was notified that her friend Mark was hospitalized with relapsed acute lymphocytic leukemia, a cancer not uncommon in persons with Down syndrome. His doctors determined that Mark was not a suitable candidate for continued treatment, and was placed in hospice care. Ann scheduled a visit, and took her infant son with her.

"Miss Ann," Mark said, his voice barely audible. "I'm so happy to see you. Is this your new baby? Is it a boy? Can I hold him?"

"Sure, Mark. You can hold him," she replied. She handed her son to Mark. "His name is Eugene," she added.

Nuzzling the child, Mark said, "Eugene — I like that name. He looks just like me. I think I will love him."

"He does, Mark. He really does look like you, and I will love him, too."

[I know you had a hand in all this — Ann's decision I mean.]

(True. I'd been talking to her for some time.)

[I, too, have been with her — from the beginning.]

(No doubt, so I expect things will work out.)

[They will. I'm certain.]